Transportation Safety Management 2018

Transportation Research Record: Journal of the Transportation Research Board Volume 2672, Issue 32

ISBN: 978-1-5108-8182-2

Printed from e-media with permission by:

Curran Associates, Inc.
57 Morehouse Lane
Red Hook, NY 12571

proceedings.com

Some format issues inherent in the e-media version may also appear in this print version.

Published by Sage Publications USA

ISBN (Print) 978-1-5108-8182-2
ISBN 2018 Printed Set (All Issues) 978-1-5108-7735-1

Printed by Curran Associates, Inc. (2019)

Additional copies of this publication are available from:

Curran Associates, Inc.
57 Morehouse Lane
Red Hook, NY 12571 USA
Phone: 845-758-0400
Fax: 845-758-2633
Email: curran@proceedings.com
Web: www.proceedings.com

Volume 2672 Number 32 December 2018

TRR
JOURNAL OF THE TRANSPORTATION RESEARCH BOARD

Contents

Transportation Safety Management 2018

Articles

Serious Road Traffic Injuries in Europe, Lessons from the EU Research Project SafetyCube 1
Wendy Weijermars, Niels Bos, Annelies Schoeters, Jean-Christophe Meunier, Nina Nuyttens, Emmanuelle Dupont, Klaus Machata, Robert Bauer, Katherine Perez, Jean-Louis Martin, Heiko Johansson, Ashleigh Filtness, Laurie Brown, and Pete Thomas

Emergency Medical Service Response: Analyzing Vehicle Dispatching Rules 10
Marco Amorim, Sara Ferreira, and António Couto

Student Pedestrian Walking Speeds at Crosswalks Near Schools 22
Kevin Chang, Peter Foss, Meagan Larrea, and Edinson Bautista

Development of a Crash Risk-Scoring Tool for Pedestrian and Bicycle Projects in Oregon 30
Yi Wang, Christopher M. Monsere, Chen Chen, and Haizhong Wang

Getting to Zero Deaths on Ohio's Low-Volume Roads 40
Victoria F. Beale, Derek Troyer, Alejandro Chock, Cory Hopwood, and Mike McNeill

Integrated Modeling Approach for Non-Motorized Mode Trips and Fatal Crashes in the Framework of Transportation Safety Planning 49
Jaeyoung Lee, Mohamed Abdel-Aty, Qing Cai, Ling Wang, and Helai Huang

Analysis of Fatal Traffic Crash-Reporting and Reporting-Arrival Time Intervals of Emergency Medical Services 61
Jaeyoung Lee, Mohamed Abdel-Aty, Qing Cai, and Ling Wang

Vehicle Consumer Complaint Reports Involving Severe Incidents: Mining Large Contingency Tables 72
Subasish Das, Abhisek Mudgal, Anandi Dutta, and Srinivas R. Geedipally

Service Location Optimization Model for Improving Rural Emergency Medical Services 83
Zhaoxiang He, Xiao Qin, Yuanchang Xie, and Jianhua Guo

Land Suitability Analysis for Emergency Medical Services Posts along State Highways: A Case Study of California 94
Ernest O. A. Tufuor, Laurence R. Rilett, Yunwoo Nam, and Antonio Hurtado Beltran

Impacts of State-Specific Policy and Legislation on Safety Advancement by Departments of Transportation 107
Jennifer H. Ogle, Sababa Islam, Kweku Brown, William J. Davis, and Wayne A. Sarasua

---

* Membership as of February 2019.

TRB

TRANSPORTATION RESEARCH BOARD

CORE SPONSORS

Transportation Departments of the 50 States and the District of Columbia

Federal Government

U.S. Department of Transportation
- Federal Aviation Administration (FAA)
- Federal Highway Administration
- Federal Motor Carrier Safety Administration (FMCSA)
- Federal Railroad Administration (FRA)
- Federal Transit Administration
- National Highway Traffic Safety Administration (NHTSA)
- Office of the Assistant Secretary of Transportation for Research and Technology (OST-R)

U.S. Air Force Civil Engineer Center (AFCEC)
U.S. Coast Guard (USCG)
U.S. Department of Energy
U.S. Department of the Interior (DOI)
U.S. Environmental Protection Agency (EPA)

Nongovernmental Organizations

American Public Transportation Association (APTA)
Association of American Railroads (AAR)
California Air Resources Board

The National Academies of

SCIENCES • ENGINEERING • MEDICINE

The **National Academy of Sciences** was established in 1863 by an Act of Congress, signed by President Lincoln, as a private, nongovernmental institution to advise the nation on issues related to science and technology. Members are elected by their peers for outstanding contributions to research. Dr. Marcia McNutt is president.

The **National Academy of Engineering** was established in 1964 under the charter of the National Academy of Sciences to bring the practices of engineering to advising the nation. Members are elected by their peers for extraordinary contributions to engineering. Dr. C. D. Mote, Jr., is president.

The **National Academy of Medicine** (formerly the Institute of Medicine) was established in 1970 under the charter of the National Academy of Sciences to advise the nation on medical and health issues. Members are elected by their peers for distinguished contributions to medicine and health. Dr. Victor J. Dzau is president.

The three Academies work together as the National Academies of Sciences, Engineering, and Medicine to provide independent, objective analysis and advice to the nation and conduct other activities to solve complex problems and inform public policy decisions. The Academies also encourage education and research, recognize outstanding contributions to knowledge, and increase public understanding in matters of science, engineering, and medicine.

Learn more about the National Academies of Sciences, Engineering, and Medicine at **www.national-academies.org.**

The **Transportation Research Board** is one of seven major programs of the National Academies of Sciences, Engineering, and Medicine. The mission of the Transportation Research Board is to increase the benefits that transportation contributes to society by providing leadership in transportation innovation and progress through research and information exchange, conducted within a setting that is objective, interdisciplinary, and multimodal. The Board's varied activities annually engage about 7,000 engineers, scientists, and other transportation researchers and practitioners from the public and private sectors and academia, all of whom contribute their expertise in the public interest. The program is supported by state transportation departments, federal agencies including the component administrations of the U.S. Department of Transportation, and other organizations and individuals interested in the development of transportation.

Learn more about the Transportation Research Board at **www.TRB.org**.

TRR
JOURNAL OF THE TRANSPORTATION RESEARCH BOARD

Article

Transportation Research Record
2018, Vol. 2672(32) 1–9

DOI: 10.1177/0361198118758055
journals.sagepub.com/home/trr
SAGE

Serious Road Traffic Injuries in Europe, Lessons from the EU Research Project SafetyCube

Wendy Weijermars[1], Niels Bos[1], Annelies Schoeters[2], Jean-Christophe Meunier[2], Nina Nuyttens[2], Emmanuelle Dupont[2], Klaus Machata[3], Robert Bauer[3], Katherine Perez[4], Jean-Louis Martin[5], Heiko Johansson[6], Ashleigh Filtness[7], Laurie Brown[7], and Pete Thomas[7]

Abstract

The EU research project SafetyCube pays specific attention to serious road injuries, defined as nonfatal road traffic casualties with an injury severity level of MAIS3+. By means of surveys, information was collected on current practices concerning the estimation of the number of MAIS3+ casualties and on costs related to serious road injuries in different European countries. Moreover, the effect of differences in practices on the estimated number of MAIS3+ casualties was investigated by applying different methods to the same data. Finally, by means of a literature review, analysis of additional case studies, and burden of injury calculations, health impacts of serious road injuries were investigated. This paper presents six main lessons learned from these activities. Practices concerning the estimation of the number of MAIS3+ casualties differ between countries; some countries apply correction factors to police data, other countries use hospital data, and a third group of countries uses linked police and hospital data. Practices also differ concerning the selection of MAIS3+ road traffic injuries within hospital data. Differences in methodology appear to affect the MAIS3+ estimate. Therefore, one should be careful when comparing figures from different countries. The SafetyCube guidelines can support further harmonization. It is important to reduce the number of serious road injuries because injuries can have major impacts on a casualty's life and pose a burden to society. About 75% of the MAIS3+ road traffic casualties are not fully recovered 3 years postcrash. Moreover, serious road injuries cost countries up to 2.7% of their gross domestic product.

According to the European Commission, an estimated 135,000 casualties were seriously injured on Europe's roads in 2015 (*1*). Worldwide, road traffic crashes result in 78.2 million nonfatal injuries warranting medical care annually, including 9.2 million injuries requiring hospital admission (*2*). According to the Global Burden of Disease Study 2013, nonfatal road traffic injuries lead to 8.6 million years lived with disability (YLD) worldwide (*3*). About 30% of the total burden of nonfatal injuries is due to road traffic injuries (*4*).

Because of their high numbers and large health impacts, serious road injuries have recently been adopted as an additional road safety performance indicator. Reducing the number of serious road injuries is one of the key priorities in the EU Road Safety Program 2011–2020 (*5*). In June 2017, the European Union transport ministers agreed to set a target of halving the number of serious injuries on EU roads between 2020 and 2030 (*6*). Furthermore, the Sustainable Development Goals of the United Nations include a target on nonfatal road injuries: the target is to halve the number of road traffic injuries between 2010 and 2020.

Serious road traffic injuries can be defined in several ways, for example as casualties admitted to a hospital, or on the basis of injury severity levels. To be able to estimate the number of serious road injuries in the EU and for comparing rates and developments between countries, it is essential to

[1]SWOV Institute of Road Safety Research, The Hague, the Netherlands
[2]VIAS Institute, Brussels, Belgium
[3]Austrian Road Safety Board KFV, Vienna, Austria
[4]Agència de Salut Pública de Barcelona, ASPB, Barcelona, Spain
[5]IFSTTAR, Cité Descartes, Champs sur Marne, France
[6]Medizinische Hochschule Hannover MHH, Hannover, Germany
[7]Loughborough Design School, Loughborough University, Loughborough, UK

Corresponding Author:
Address correspondence to Wendy Weijermars:
Wendy.weijermars@swov.nl

SafetyCube project
SafetyCube (Safety CaUsation, Benefits and Efficiency) is an EU - funded research project that started in May 2015 and will be finished in April 2018.
Within SafetyCube an **innovative road safety Decision Support System (DSS)** is developed, that aims to enable road policy makers and other stakeholders to select the most appropriate measures to reduce casualties of all road user types and all severities in Europe and worldwide. The DSS provides information about a wide variety of behaviour, infrastructure and vehicle - related risk factors and about the effectiveness and cost-effectiveness of measures that can be taken to reduce these risk factors. The DSS is available through: roadsafety-dss.eu

Figure 1. General information about the EU funded SafetyCube project.

have a common definition. In 2013 a common definition was established by the EU member states (*7*). Serious road injuries are defined as nonfatal road traffic casualties with an injury severity level of MAIS3+ (Maximum Abbreviated Injury Scale) (*8*). However, methodologies for estimating the number of serious road injuries differ between countries and it is not known to what extent differences in methodology influence the estimated number of MAIS3+ casualties.

As serious road traffic injuries are a relatively new road safety performance indicator, information on the numbers, health impacts, and costs related to serious road traffic injuries is quite scarce. Therefore, one of the work packages of the EU research project SafetyCube (hsafetycube-project.eu/) looked into:

- Current practices concerning the estimation of the number of serious road injuries across Europe and consequences of methodological differences for the estimated number;
- Health impacts of serious road traffic injuries;
- Costs related to serious road traffic injuries; and
- Risk factors that are associated with serious road traffic injuries.

This paper discusses the current practices concerning the estimation of the number of serious road injuries, as well as the health impacts and costs related to serious road injuries. Figure 1 provides more information on the SafetyCube project in general.

Method

Determination of Current Practices and Effect of Method on Estimated Number

The High Level Group on Road Safety – based in Brussels, and convening all EU Member States – identified three main ways in which Member States can collect data on MAIS3+ road traffic casualties:

1. By applying a correction to police data,
2. By using hospital data, and
3. By using linked police and hospital data.

Within each method, additional choices need to be made, for example concerning the derivation of correction factors that are applied to police data, the selection of road traffic injuries in hospital data, and the selection of casualties with an injury severity of MAIS3+. The choices largely depend on the available data and may affect the estimated number of serious road injuries.

An expert survey of 32 European countries was conducted, documenting current and planned practices concerning the estimation of the number of MAIS3+ road traffic injuries. The questionnaire was inspired by a survey that had been conducted by the Forum of European Road Safety Research Institutes (FERSI) (*9*). The response rate was over 80%: 26 out of 32 countries responded to the questionnaire.

To examine the comparability of numbers determined by different methods, all three methods proposed by the High Level Group were applied to the Netherlands. The "official" estimated number of serious road injuries in the Netherlands is based on linked police and hospital data, including a "capture–recapture" estimate of casualties that are missing in both datasets. Within SafetyCube, this estimate was compared with the estimate that results from applying correction factors to police data and the estimate that results from using hospital data only. A set of 70 correction factors for different combinations of age, gender, and transport mode was derived using police and hospital data for the period 2004–2008. To examine whether correction factors are stable over time, the set of correction factors determined for the period 2004–2008 was also applied to estimate the number of MAIS3+ casualties in the period 2009–2014.

In addition to the comparison of the three main methods proposed by the High Level Group discussed in this paper, within SafetyCube, the consequences of differences in methodology between countries that applied hospital data were also analyzed. These differences are mainly related to (*a*) the selection of road traffic casualties within all injury patients, and (*b*) the selection of casualties with injury severity MAIS3+. The consequences of differences were analyzed by applying different selection criteria to the same data. These sensitivity analyses, using German, Dutch, Spanish, and Belgium data are discussed in more detail in Pérez et al. (*10, 11*).

Health Impacts of Serious Road Traffic Injuries

Road traffic injuries can have a variety of consequences for individual casualties, for their relatives and friends, for other people involved in the crash, and for society as a whole. Within SafetyCube, the consequences for individual casualties, and for society as a whole were the focus.

Consequences for individual casualties were investigated by means of a literature review and analysis of several additional studies and data that project partners had access to. One of the additional studies included is the ESPARR study (e.g., *12, 13*), a prospective cohort study that follows a group of more than 1,100 road traffic casualties for up to 5 years, including 320 MAIS3+ casualties, seeking medical care in health facilities in the Rhône administrative area in France.

According to the International Classification of Functioning, Disability and Health (ICF), injuries may result in disabilities related to one or more levels of human functioning (*14*):

- Impairments, that is, problems related to body function or body structure, for example paralysis;
- Activity limitations, for example being unable to walk; and
- Participation restrictions, such as being unable to work.

The extent to which an injury affects human functioning also depends on personal and environmental factors. The literature review and analysis of additional data/studies included effects on all three levels of human functioning, as well as the influence of personal and environmental factors.

The burden of nonfatal injuries to society can be expressed by disability adjusted life years. This measure integrates mortality – expressed in years of life lost (YLL) – and morbidity, expressed in YLD and has been applied to road traffic injuries before (e.g., *2, 15, 16*). Within SafetyCube, the method that was developed within the European INTEGRIS study (*17*) was applied to calculate the burden of nonfatal MAIS3+ road traffic injuries. The method combines incidence data of injuries with disability information for these injuries, applying the EUROCOST injury classification (*18*). The burden of injury was calculated for the following countries/regions: Austria, Belgium, England, the Netherlands, the Rhône department of France, and Spain. For more information on the burden of injury calculations done in SafetyCube see Weijermars et al. (*19, 20*).

Costs Related to Serious Road Traffic Injuries

On the basis of a survey among 32 European countries, crash cost estimates were collected for 30 European countries. The data collection was a joint effort between the EU projects SafetyCube and InDeV (http://www.indev-project.eu/). The questionnaire included questions concerning available information on costs and methodologies applied for determining these costs, as well as questions regarding cost estimates per crash and casualty by severity, estimates per cost component by severity, and the total costs of crashes and their percentage of GDP. The questionnaires were prefilled by project partners using available crash cost reports and were then sent to a selected expert in each country for checking and completion. Within SafetyCube, the questionnaires were integrated into a SQLite database and multiple consistency checks were carried out, resulting in several corrections. To be able to compare costs from different countries, all values are expressed in EUR price level 2015 and adjusted for relative income differences using Purchasing Power Parities (PPP). PPP are the rates of currency conversion that equalize the purchasing power of different currencies, they are price relatives that show the ratio of the prices in national currencies of the same good or service in different countries (*21*). For more information on the data collection, see Wijnen et al. (*22*).

The following six cost components were distinguished in the survey, on the basis of a review of international guidelines and best practices (*22*):

1. Medical costs;
2. Costs related to production loss;
3. Human costs: costs related to loss of quality of life, grief, pain, sorrow;
4. Administrative costs, for example police and fire brigade costs related to the handling of road crashes;
5. Costs related to property damage;
6. Other costs, for instance congestion costs.

In relation to serious road injuries, medical costs, costs related to production loss and human costs are most relevant (*23*). Furthermore, crashes with serious road injuries also induce crash-related costs including administrative costs, property damage and other costs.

Results

Current Practices and Effect of Method on Estimated Number

In June 2016, 17 out of the 26 countries that responded to the survey had either already estimated the number of MAIS3+ casualties or reported that they would be able to estimate the number very soon. The remaining nine countries were in a very early stage of the estimation process or had not yet started. One of the main reported problems was accessing hospital discharge data; this was due to privacy regulations.

The methods for estimating the number of MAIS3+ road traffic casualties appear to differ between countries. Two countries applied correction factors to police data, nine countries used only hospital data, four countries used linked police and hospital data, and two countries applied a combination of methods. Several countries mentioned plans for modifying

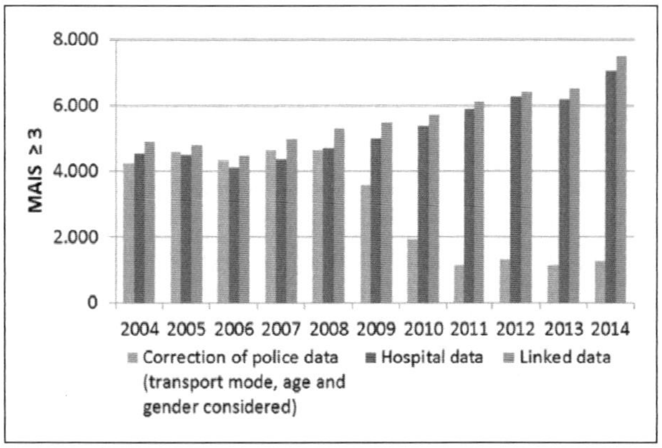

Figure 2. Comparison of number of MAIS3+ casualties per year for three methods in the Netherlands for 2004–2014.

their method in the future, the majority of them toward linking police and hospital data.

The survey also revealed that the nine countries that used hospital data, show differences concerning the selection of road traffic casualties and the determination of the MAIS level of casualties. In some countries it appeared to be problematic to identify road traffic casualties within all injury patients. In Poland for example, for 38% of the injury patients in the hospital data, the cause of the injury is missing. In addition, some countries experience difficulties in determining the MAIS level of injury patients. In Austria for example, for about 20% of the injury patients, the MAIS could not be determined. The sensitivity analyses discussed in Pérez et al. (*10, 11*) show that differences in hospital data can lead to substantial differences in the estimated number of MAIS3+ casualties.

Figure 2 shows the estimated numbers of serious road injuries in the Netherlands resulting from the three different methods. The figure shows that for all years, the estimated number of MAIS3+ casualties is highest when using linked police and hospital data. This can be explained by the linking procedure that is applied, which includes an estimation of the number of casualties that is missing in both police and hospital data. Police data deals with quite high levels of underreporting in the Netherlands: in 2004, 41% of the MAIS3+ casualties were reported by the police and in 2014 this was only 24%. The 70 correction factors that were estimated within SafetyCube appear unable to fully correct for the underreporting by the police. All MAIS3+ road traffic casualties are assumed to be reported in the hospital data, but as a result of missing or incorrect classification of external causes, they are not always recognized as a road traffic casualty. As a consequence, the estimated number of MAIS3+ casualties on the basis of only hospital data is lower than the estimate based on linking police and hospital data.

Figure 2 also shows that applying correction factors to police data for 2009 onwards, results in a very high underestimation of the number of MAIS3+ casualties. This can be explained by the dramatic decrease in the number of police-reported casualties since 2009, due to new regulations, and problems with a new reporting system. This example shows that one should be careful when applying correction factors to police data. Before correction factors are applied, one should check whether registration levels remained constant.

Health Impacts of Serious Road Traffic Injuries

Both the literature and the additional studies show that non-fatal road traffic injuries can have a major impact on the lives of casualties (and their families). Reported prevalence of disabilities varies widely between studies, depending on the casualties included in the study (e.g. injury severity levels), the duration of follow-up and the type of disabilities taken into account. According to a recent literature review (*16*), self-reported prevalence of disability resulting from road traffic injuries varies between 11% and 80%. Data from the ESPARR cohort study shows that about 75% of the MAIS3+ road traffic casualties have not fully recovered 3 years after the crash (*19*).

Reported consequences of road traffic injuries relate to all three levels of human functioning defined in the ICF and include for example pain, fatigue, mobility problems, sick leave and problems carrying out daily activities. Additionally, road traffic injuries lead to psychological disorders like Post Traumatic Stress Disorder (PTSD) and fear, which in their turn also affect human functioning. Literature reports a prevalence of PTSD between 2% and 33% 1 year postcrash (*19*). Moreover, the problems related to human functioning can also lead to socioeconomic consequences like financial or relational problems. Data from the ESPARR study for example shows that more than 10% of MAIS3+ casualties encounter financial difficulties 5 years postcrash (*19*).

The burden of injury calculations reveal that the average burden of injury per MAIS3+ casualty varies between 2.4 YLD in Spain and 3.2 YLD in the Netherlands. Moreover, between 19% (Spain) and 33% (the Netherlands) of the MAIS3+ casualties experience lifelong disabilities. These lifelong disabilities are responsible for about 90% of the total burden of injury of MAIS3+ casualties. For the Netherlands and England, the burden of injury of MAIS3+ casualties (expressed in YLD) was compared with the burden of injury of fatalities (expressed in YLL). In the Netherlands, the burden of injury of MAIS3+ casualties appeared to be similar to the burden of fatalities.

Reported consequences differ between casualties, depending on for example the injury sustained, age, comorbidity, and socioeconomic status. The average burden of injury per casualty depends on the injury and the age of the casualty. According to the literature, comorbidity and a lower socioeconomic status are associated with a higher risk of experiencing consequences. Concerning age, younger casualties have a lower risk of encountering physical consequences, but have a higher average burden per injury as they have a longer

Table 1. Summary Information Concerning the Burden of Injury Per Hospitalized Casualty for Different Severity Levels and the Percentage of MAIS3+ in the Total Number and Burden of Injury of Hospitalized Casualties

Country	Burden per person			MAIS3+ as % of hospitalized	Burden MAIS3+ as % of burden hospitalized
	MAIS3+	MAIS2	MAIS1		
Austria	3.1	1.0		9%	22%
Belgium	2.7	1.6	0.6	22%	44%
England	3.1	2.1	0.3	17%	34%
Netherlands	3.2	1.6	0.5	26%	48%
Rhône	2.5	1.9	0.6	45%	58%
Spain	2.4	1.3		34%	48%

remaining life. Concerning the injury sustained, studies quite consistently show that consequences increase with injury severity, although minor injuries, like whiplash-associated disorders, may also have substantial long-term consequences. This can also be seen from the burden of injury calculations. However, as less severe injuries are much more common, they constitute quite a high share of the total burden of injury in a country. Table 1 shows information on the burden of injury for hospitalized casualties with severity levels MAIS1, MAIS2, and MAIS3+ in various countries. The Netherlands and the Rhône department in France also have information on the burden of injury of road traffic injuries that are treated at an emergency department and discharged to the home environment. For the Netherlands, Polinder et al. estimated that casualties treated at the Emergency Department, are responsible for 26% of the total burden of nonfatal injury (*24*). Therefore, MAIS3+ casualties are responsible for only 36% (48%*74%) of the total burden of nonfatal traffic injuries in the Netherlands. For the Rhône department, almost 7% of all nonfatal road traffic casualties (including casualties treated at an emergency department or admitted to a hospital) have an injury severity of MAIS3+. These MAIS3+ casualties are responsible for 26% of the total burden of injury of all treated nonfatal traffic casualties.

Costs Related to Serious Road Traffic Injuries

The survey showed that costs related to serious road injuries differ considerably between European countries. Figure 3 shows that the costs per serious road injury vary between roughly €28,000 and €975,000 per casualty. Figure 4 shows that the total costs related to serious injuries vary between 0.04% and 2.7% of a country's GDP. Costs related to serious road injuries account for 14% to 77% of the total costs of road crashes in a country. There seem to be a couple of outliers in figures 3 and 4: in Poland and in Estonia, the costs per serious road injury (Figure 3) and the total costs of serious road injuries (Figure 4) appear to be very high. In addition, total costs are relatively high in Hungary and Croatia. It should be noted that information on costs is provided by experts, and for some countries – especially Poland and Hungary – there is very little information on how these costs are calculated. Further researched is needed into costs of serious road injuries in these countries.

Differences between the other countries are mainly due to whether or not the Willingness to Pay (WTP) method is applied for calculating the human costs. In a WTP study, the amount of money an individual or society is willing to pay for a risk reduction is estimated, and from these results, the so called Value of a Statistical Life (VOSL) is derived. The VOSL is subsequently used to calculate human costs of fatalities. Information on human costs of serious injuries is relatively poor compared with human costs of fatalities, although in some countries thorough WTP studies are carried out in which the WTP for reducing the risk of getting injured is estimated relative to the WTP for reducing the risk of being killed (*25, 26*). The WTP method is the generally recommended method for the calculation of human costs (*26, 27*) and results in higher human costs than alternative approaches such as the use of financial compensations that are awarded to road casualties in courts of law that is applied in Germany (*28*) and Australia (*29*).

Other factors that contribute to differences between countries are differences in the definition of a serious road injury, differences regarding the cost components that are included, and differences in reporting rates of serious road injuries. In addition, differences in total costs of serious road injuries in relation to a country's GDP are also due to differences in road safety levels.

In most countries, human costs represent the largest share in the costs of serious road injuries. This is particularly the case for countries that use the WTP method; in these countries the share of human costs varies between 51% and 91%. However, some countries show a high proportion of human costs because other cost components are not included. This applies particularly to countries that use the HEATCO approach (*26*), which implies that all costs other than human costs and consumption loss are estimated at 10% of the VOSL. Medical costs and production loss are the two other main components of the costs related to serious injuries. These components both have a median share in the total costs of 18% (only taking into account countries that included these components).

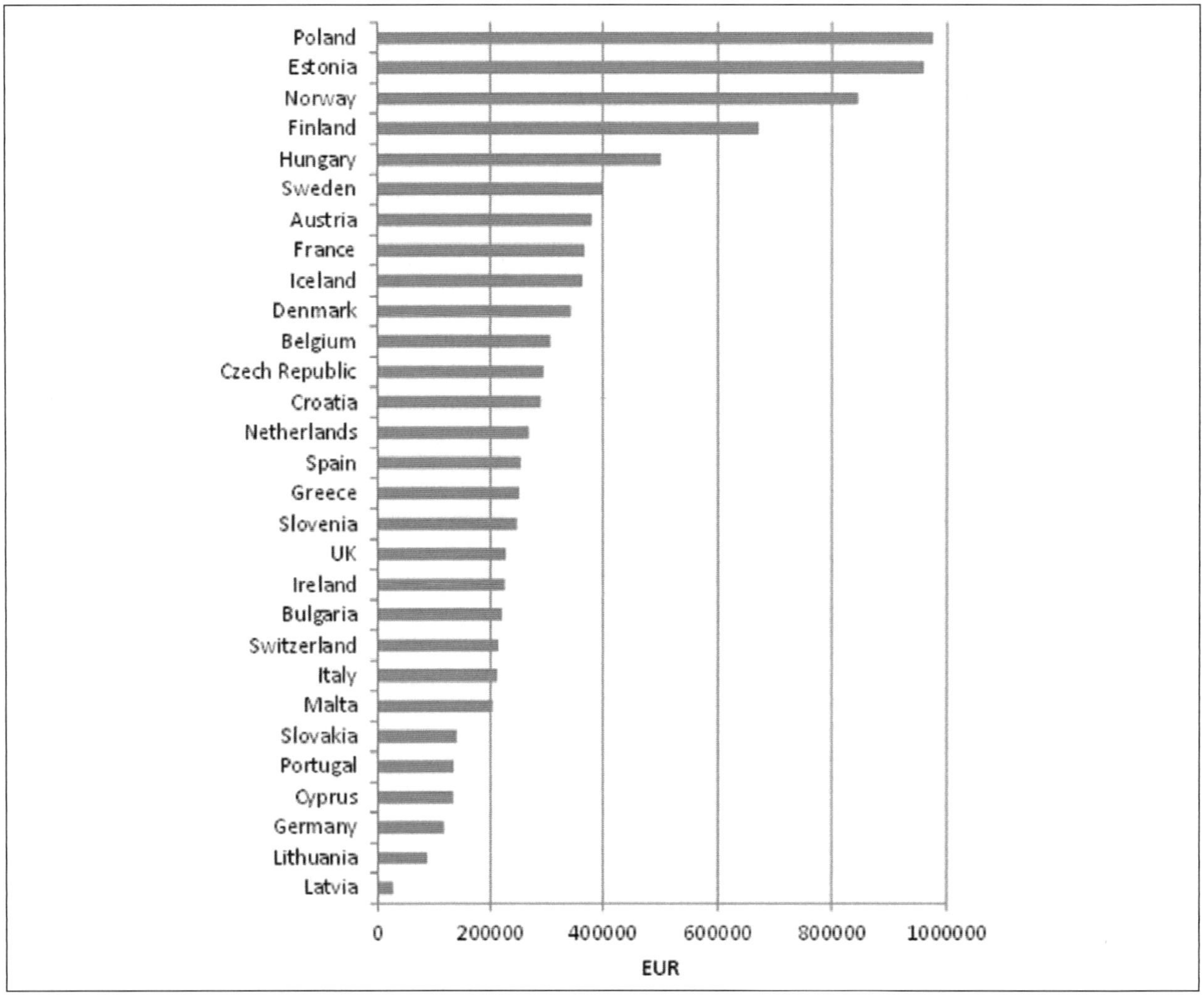

Figure 3. Costs per serious road injury (EUR 2015, adjusted for PPP) (22).

Main Lessons Learned from SafetyCube

This section summarizes the lessons that were learned from the work undertaken within SafetyCube on serious road injuries.

1. **As practices concerning the estimation of the number of MAIS3+ casualties differ between countries, and differences in methodology affect the estimate, one should be careful when comparing figures from different countries.**
   From the survey it can be concluded that practices differ between countries, depending on the available data. Moreover, (sensitivity) analyses showed that these differences have a considerable effect on the estimated number of MAIS3+ casualties. Therefore, one should be careful when comparing figures from different countries. Differences in methodology should be reported and discussed. Moreover, further harmonization of methods is certainly desirable to ensure that the estimated numbers of MAIS3+ casualties are comparable across Europe. The guidelines that are developed within SafetyCube can help improve comparability. A four-page summary of the guidelines can be found at: https://www.safetycube-project.eu/wp-content/uploads/SafetyCube-D7.1-Leaflet.pdf
2. **When applying correction to police data to estimate the number of MAIS3+ casualties, it is important to assure police registration levels are stable.**
   From the comparison of the three methods for calculating MAIS 3+ using data from the Netherlands it can be seen that application of correction factors to police data can result in a huge underestimation when

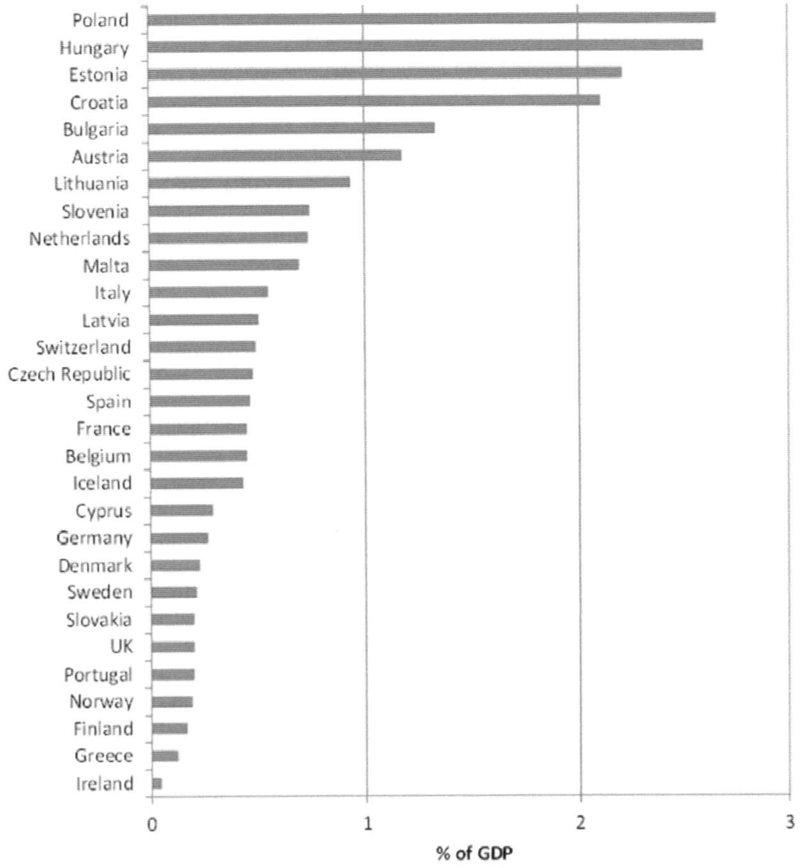

Figure 4. Total costs related to serious road injuries as a percentage of GDP (22).

reporting rate drops. Therefore, one should be careful when applying correction factors to police data. One should ensure that police reporting rates are stable, and one should have access to at least a sample of good-quality hospital data in order to estimate reliable correction factors.

3. **Good-quality hospital data is essential for the estimation of the number of MAIS3+ casualties.**
All three methods proposed by the EU High Level Group on road safety require at least a sample of good-quality hospital data. The survey revealed that some EU countries encounter difficulties accessing hospital discharge data, due to privacy regulations. Furthermore, some countries experience problems related to the quality of hospital data that complicate the estimation of the number of MAIS3+ road traffic injuries. More intersectorial collaboration between the health and the transport actors at national and international level could be beneficial for improving hospital data. Furthermore, correction factors might be applied to hospital data to provide a more accurate estimate of the number of MAIS3+ casualties and linking of police and hospital data might be beneficial in case of problems related to the selection of road traffic casualties in hospital data.

4. **It is important to reduce the number of MAIS3+ casualties as AIS3+ injuries can have major impacts on casualties' lives, pose a burden to society and result in considerable costs.**
From the literature review on consequences of road traffic injuries and the additional studies can be concluded that road traffic injuries have long-term consequences for quite a high proportion of MAIS3+ casualties. The ESPARR study for example shows that about 75% of the MAIS3+ road traffic casualties have not fully recovered 3 years after the crash. The burden of injury calculations show that a MAIS3+ casualty, on average has a burden of 2.4 to 3.2 YLD. In total 19% to 33% of the MAIS3+ casualties experience lifelong disabilities. In addition, in the Netherlands, the total health burden of MAIS3+ casualties is comparable to the health burden of fatalities. Finally, cost estimates show that serious road injuries cost countries up to 2.7% of their GDP each

year and they account for up to 77% of the total costs of road crashes in a country.

5. **From a burden of injury perspective, less serious injuries are also relevant.**
   Although, in general, less seriously injured casualties experience long-term consequences less often, consequences can still be severe for minor injuries. Moreover, because of their high number, they are also relevant from a burden of injury perspective. More than half of the burden of nonfatal road traffic injuries appears to be due to MAIS2- injuries (including injuries that are only treated at an emergency department).
6. **Cost estimates differ considerably between countries.**
   The cost estimates collected by the survey appeared to differ considerably between countries. Costs per serious road injury for example vary between €28,000 and €975,000 per casualty. Differences between countries are mainly due to whether or not the WTP method is applied for calculating the human costs.

Conclusion

This paper presents the main lessons learned from the SafetyCube research on serious road injuries. On an EU level, serious road injuries are defined as nonfatal road traffic casualties with an injury severity of MAIS3+. EU Member States collect data on serious road injuries by applying one of the following methods: 1) by applying a correction to police data, 2) by using hospital data, and 3) by using linked police and hospital data.

The methodology applied for estimating the number of MAIS3+ casualties appears to differ between countries, mainly as a result of differences in data availability. The differences in methodology appear to have a considerable effect on the MAIS3+ estimate; therefore, one should be careful when comparing figures from different countries. The guidelines that were developed in SafetyCube aim to improve comparability of estimates from different countries.

The literature review and additional case studies show that serious road injuries can have significant consequences for individual casualties. About 75% of the MAIS3+ casualties report that they are not fully recovered 5 years postcrash. The burden of injury calculations show that serious road injuries also pose a considerable burden to society. In the Netherlands, the health burden of MAIS3+ road casualties is comparable to the health burden of road fatalities. Cost estimates appear to differ between countries. Serious road injuries cost countries up to 2.7% of their GDP each year, and costs of serious road injuries account for up to 77% of the total costs of road crashes in a country. Because of their high costs and potentially severe consequences, road safety policy should also be aimed at reducing the number of (serious) nonfatal road injuries, next to reducing the number of fatalities.

Acknowledgments

This paper is based on work carried out within the SafetyCube research project of the H2020 program of the European Commission (Grant no. 633485). Moreover, the work is based on analyses carried out by the following other persons (next to the authors): M. Olabarria, X. Duran (ASPB), S. Houwing, G. J. Wijlhuizen, W. Wijnen, (SWOV), E. Amoros, L. Pascal, M. Hours, L. Carnis, (IFSTTAR), W. Van den Berghe, S. Daniels (BRSI), B. Leskovsek (AVP), and R. Elvik (TOI).

References

1. European Commission press release. *Road Safety: New Statistics call for Fresh Efforts to Save Lives on EU Roads*. http://europa.eu/rapid/press-release_IP-16-863_en.htm. Accessed November 10, 2017.
2. Bhalla, K., M. Shotten, A. Cohen, M. Brauer, S. Shahraz, R. Burnett, K. Leach-Kemon, G. Freedman, and C. J. Murray. *Transport for Health: The Global Burden of Disease from Motorized Road Transport*. World Bank Group, London, UK, 2014.
3. Haagsma, J. A., N. Graetz, I. Bolliger, M. Naghavi, H. Higashi, E.C. Mullany, S. F. Abera, et al. The Global Burden of Injury: Incidence, Mortality, Disability-Adjusted Life Years and Time Trends from the Global Burden of Disease Study 2013. *Injury Prevention*, Vol. 22, No. 1, 2016, pp. 3–18. http://doi.org/10.1136/injuryprev-2015-041616
4. Vos, T., A. Flaxman, M. Naghavi, R. Lozano, ... C. Murray. Years lived with disability (YLDs) for 1160 Sequelae of 289 Diseases and Injuries 1990–2010: A Systematic Analysis for the Global Burden of Disease Study 2010. *The Lancet*, Vol. 380, 2012, pp. 2163–2196.
5. European Commission. *Communication from the Commission to the European Parliament, the Council, the European Economic and Social Committee and the Committee of the Regions: Towards a European Road Safety Area: Policy Orientations on Road Safety 2011–2020*. European Commission, Brussels, 2010.
6. European Transport Safety Council (ETSC) press release. *EU Sets New Target to Cut Serious Road Injuries*. http://etsc.eu/eu-sets-new-target-to-cut-serious-road-injuries/. Accessed November 10, 2017.
7. European Commission. *Commission Staff Working Document on the Implementation of Objective 6 of the European Commission's Policy Orientations on Road Safety 2011–2020: First Milestone Towards an Injury Strategy*. SWD(2013) 94 final. European Commission, Brussels, 2013.
8. Gennarelli, T. A., and E. Wodzin. *Abbreviated Injury Scale (AIS) 2005* [Manual]. Association for the Advancement of Automotive Medicine, Chicago, Ill., 2005.
9. Auerbach, K., and U. Schmucker. *Country Survey State of the Art of MAIS 3+ Assessment in the FERSI Member States and EU/EEA countries*. FERSI working group "Injury Classification," 2016.
10. Pérez, K., W. Weijermars, E. Amoros, R. Bauer, N. Bos, E. Dupont, A. Filtness, S. Houwing, H. Johannsen, B. Leskovsek, K. Machata, J. L. Martin, N. Nuyttens, M. Olabarria, L. Pascal, and

W. Van den Berghe. *Practical Guidelines for the Registration and Monitoring of Serious Traffic Injuries*, D7.1 of the H2020 project SafetyCube. Loughborough University,Leicestershire, UK, 2016.

11. Perez, K, W. Weijermars, N. Bos, A. Filtness, R. Bauer, H. Johanssen, N. Nuyttens, L. Pascal, and P. Thomas. Implications of Estimating Road Traffic Serious Injuries from Hospital Data. *Submitted to Accident Analysis & Prevention*, forthcoming.
12. Hours, M., M. Bernard, P. Charnay, L. Chossegros, E. Javouhey, E. Fort, D. Boisson, P. O. Sancho, and B. Laumon. Functional Outcome after Road-Crash Injury: Description of the ESPARR Victims Cohort and 6-Month Follow-Up Results. *Accident Analysis & Prevention*, Vol. 42, No. 2, 2010, pp. 412–421. http://doi.org/10.1016/j.aap.2009.09.002
13. Hours, M., L. Chossegros, P. Charnay, H. Tardy, H-T. Nhac-Vu, D. Boisson, and B. Laumon. Outcomes One Year after a Road Accident: Results from the ESPARR cohort. *Accident Analysis & Prevention*, Vol. 50, 2013, pp. 92–102.
14. World Health Organization. *Towards a Common Language for Functioning, Disability and Health, ICF*. World Health Organization, Geneva, Switzerland, 2002.
15. Dhondt, S., C. Macharis, N. Terryn, F. Van Malderen, and K. Putman. Health Burden of Road Traffic Accidents, an Analysis of Clinical Data on Disability and Mortality Exposure Rates in Flanders and Brussels. *Accident Analysis & Prevention*. Vol. 50, 2013, pp. 659–666. http://doi.org/http://dx.doi.org/10.1016/j.aap.2012.06.019
16. Weijermars, W., N. Bos, and H. Stipdonk. Health Burden of Serious Road Injuries in the Netherlands. *Traffic Injury Prevention*. Vol. 17, No. 8, 2016, pp. 863–869. http://doi.org/10.1080/15389588.2016.1157591
17. Haagsma, J. A., S. Polinder, R.A. Lyons, J. Lund, V. Ditsuwan, M. Prinsloo, J. L. Veerman, and E.F. van Beeck. Improved and Standardized Method for Assessing Years Lived with Disability after Injury. *Bulletin of the World Health Organisation*, Vol. 90, No. 7, 2012, pp. 513–521. http://doi.org/10.2471/blt.11.095109
18. Polinder, S., W. Meerding, H. Toet, M. van Baar, S. Mulder, and E. van Beeck. *A Surveillance Based Assessment of Medical Costs of Injury in Europe: Phase 2*. Consumer Safety Institute, Amsterdam, the Netherlands, 2004.
19. Weijermars, W., J.-C. Meunier, N. Bos, C. Perez, M. Hours, H. Johannsen, J. Barnes, G.-J. Wijlhuizen, N. Nuytens, E. Dupont, L. Brown, C. Quigley, A. Filtness, X. Duran, J. Martin, and R. Bauer. *Physical and Psychological Consequences of Serious Road Traffic Injuries*. D7.2 of the H2020 project SafetyCube. Loughborough University, Leicestershire, UK, 2016.
20. Weijermars, W., N. Bos, A. Filtness, L. Brown, R. Bauer, E. Dupont, JL. Martin, K. Perez, and P. Thomas. Burden of injury of serious road injuries in six EU countries. *Accident Analysis & Prevention*. Vol 111, 2018, pp. 184-192. https://doi.org/10.1016/j.aap.2017.11.040.
21. *Eurostat-OECD Methodological Manual on Purchasing Power Parities*. Eurostat/OECD, Publications Office of the European Union, Luxembourg, 2012.
22. Wijnen, W., W. Weijermars, W. Vanden Berghe, A. Schoeters, R. Bauer, L. Carnis, R. Elvik, A. Theofilatos, A. Filtness, S. Reed, C. Perez, and H. Martensen. *Crash Cost Estimates for European Countries*, D3.2 of the H2020 project SafetyCube. Loughborough University, Leicestershire, UK, 2017.
23. Schoeters, A., W. Wijnen, L. Carnis, W. Weijermars, R. Elvik, H. Johannsen, W. Vanden Berghe, A. Filtness, and S. Daniels. *Costs Related to Serious Road Injuries*, D7.3 of the H2020 project SafetyCube, Loughborough University, Leicestershire, UK, 2017.
24. Polinder, S., J. Haagsma, N. Bos, M. Panneman, K. K. Wolt, M. Brugmans, W. Weijermars, and E. Van Beeck. Burden of Road Traffic Injuries: Disability-Adjusted Life Years in Relation to Hospitalization and the Maximum Abbreviated Injury Scale. *Accident Analysis & Prevention*. Vol. 80, 2015, pp. 193–200.
25. European Conference of Ministers of Transport (ECMT). *Efficient Transport for Europe: Policies for Internalisation of External Costs*. ECMT: Distributed by the OECD Publications Service, Paris, France, 1998.
26. Bickel, P., R. Friedrich, A. Burgess, P. Fagiani, A. Hunt, G. De Jong, J. Laird, C. Lieb, G. Lindberg, P. Mackie, S. Navrud, T. Odgaard, A. Ricci, J. Shires, and L. Tavasszy. *Proposal for Harmonised Guidelines*. Deliverable 5 of the EU- HEATCO project. IER, Germany, 2006.
27. Alfaro, J. L., M. Chapuis, and F. Fabre. *Socio-Economic Cost of Road Accidents*: Final Report of Action COST 313. Commission of the European Community, Brussels, 1994.
28. Baum, H., T. Kranz, and U. Westerkamp. *Volkswirtschaftliche Kosten durch Straßenverkehrsunfälle in Deutschland. Heft M208*. Bundesanstalt für Straßenwesen, Bergisch Gladbach, Germany, 2007.
29. Bureau of Infrastructure, Transport and Regional Economics (BITRE). *Road Crash Costs in Australia 2006*. Research report 118. Department of Infrastructure, Transport, Regional Development and Local Government, Australian Government, Canberra, Australia, 2009.

The Standing Committee on Transportation Safety Management (ANB10) peer-reviewed this paper (18-00898).

The information and views set out in this paper are those of the authors and may not reflect the official opinion of the European Commission.

TRR
JOURNAL OF THE TRANSPORTATION RESEARCH BOARD

Article

Transportation Research Record
2018, Vol. 2672(32) 10–21

DOI: 10.1177/0361198118781645
journals.sagepub.com/home/trr
SAGE

Emergency Medical Service Response: Analyzing Vehicle Dispatching Rules

Marco Amorim[1], Sara Ferreira[1], and António Couto[1]

Abstract

In an era of information and advanced computing power, emergency medical services (EMS) still rely on rudimentary vehicle dispatching and reallocation rules. In many countries, road conditions such as traffic or road blocks, exact vehicle positions, and demand prediction are valuable information that is not considered when locating and dispatching emergency vehicles. Within this context, this paper presents an investigation of different EMS vehicle dispatching rules by comparing them using various metrics and frameworks. An intelligent dispatching algorithm is proposed, and survival metrics are introduced to compare the new concepts with the classic ones. This work shows that the closest idle vehicle rule (classic dispatching rule) is far from optimal and even a random dispatching of vehicles can outperform it. The proposed intelligent algorithm has the best performance in all the tested situations where resources are adequate. If resources are scarce, especially during peaks in demand, dispatching delays will occur, degrading the system's performance. In this case, no conclusion could be drawn as to which rule might be the best option. Nevertheless, it draws attention to the need for research focused on managing dispatch delays by prioritizing the waiting calls that inflict the higher penalty on the system performance. Finally, the authors conclude that the use of real traffic information introduces a considerable gain to the EMS response performance.

Emergency medical services (EMS) are a vital part of today's modern cities. These are services that must respond promptly to medical emergencies, which may occur at any time and any place. Depending on the country, the EMS "modus operandi" may vary, but in general, it relies on an emergency phone line (e.g., 911 in the USA, 112 in the EU) that directs all the medical emergencies to an EMS call center. The operator acquires the relevant information from the caller and makes a pre-assessment of the medical emergency needs. Usually, the EMS call center has an implemented algorithm that manually or automatically defines the priority of the emergency and triggers a request for a vehicle to be dispatched. The most frequently used EMS dispatching rule is to send the closest idle vehicle.

Nevertheless, it is important to investigate why and if the closest vehicle is really the best practice. One might conclude that, individually, the faster the patient is attended, the better. However, sending that vehicle might debilitate the system's response for the next emergency compared with sending a different one. This is particularly true if we take into account the type and severity of the case, where the response time may affect the patient's outcome.

Moreover, the urban environment where the system is implemented also influences the quality of the response it can provide. In fact, time-dependent information within this urban environment plays an important role in the system's response, as shown by Schmid (*1*). With recent technological improvements, practitioners and researchers are in a position to collect and integrate, in real time, large amounts of diverse data, such as traffic congestion and real-time emergency calls (*2*).

One could claim that dispatching an emergency vehicle is a complex decision that goes beyond the request for the closest unit available. It requires the management of a vital network of vehicles, the assessment of the severity and heterogeneity of the emergency cases, and an understanding of the urban environment where the system is integrated.

The objective of this study is to investigate this claim by assessing several dispatching rules in different temporal contexts. This is achieved by using various performance metrics to evaluate the versatility of the closest vehicle dispatching rule and its possible substitutes. A simulation model and a dispatching rule are

[1]Research Centre for Territory, Transports and Environment, Faculty of Engineering, University of Porto, Porto, Portugal

Corresponding Author:
Address correspondence to Marco Amorim: mraul@fe.up.pt

proposed by the authors; different performance metrics are applied and the performance of vehicle dispatching rules is assessed in different contexts with the objective of providing practitioners with a proof of concept, applicability, and empirical results in different emergency contexts.

Literature Review

The classic emergency service problem can be traced back to 1955, with the fire station location planning study by Valinsky (*3*). Additionally, Hogg (*4*) along with Savas (*5*) completed the base archetypes for this field of investigation, where the latter gives special focus to the EMS. EMS station and vehicle locations, allocation and dispatching have since been widely studied; however, there has always been a gap between theory and practice, owing to the lack of possibilities to test these models in a real environment.

With computers becoming ever more powerful and accessible to everyone, simulation is also becoming an interesting tool for researchers to formulate more realistic and complex models, either to assess solutions or to support optimization models (*6–11*). Haghani and Yang (*12*) propose a simulation model to assess EMS performance. To abstract from the complexity of the road system, they adopt a nodal network and assume that real-time traffic information is known. McCormack and Coates (*9*) also use simulation to assess vehicle allocation performance with a focus on increasing victims' survival. The use of simulation has proven to be valuable owing to the ability of directly using real EMS call data, compared with other methods where demand must be modeled, and thus simplified.

When assessing EMS performance, researchers fall back on metrics from the two most relevant works on EMS, those of Toregas et al. (*13*) and Church and Velle (*14*). The former presents a solution to the location set covering problem, guaranteeing that all demand is covered within a maximum time or distance radius. Church and Velle (*14*) approach the problem with a solution for the maximal coverage location problem, which intends to overcome the resource limitations neglected by Toregas et al. (*13*). Nevertheless, once a facility is called into service, its allocated demand points are no longer covered. Daskin and Stern (*15*, *16*) and Hogan and ReVelle (*17*, *18*) tackle this problem by adding facility busy probability and reliability.

In these models, the metric used to evaluate the system performance and/or optimize it is the response time, which is usually simplified by a maximum response time threshold, for example, percentage of population covered within a response radius of 8 min (*19*). Although it is obvious that a quicker medical response will always result in improved medical assistance (*20*, *21*), the response time affects different types of medical emergencies in different ways. For instance, Sánchez-Mangas, et al. (*22*) indicated that a response time reduction of 10 min could result in a 30% reduction of fatalities in traffic crashes, whereas Valenzuela et al. (*23*) showed that cardiac arrest fatalities could be reduced by 50% if the victims were assisted no later than 3 min after a collapse. Thus, by relying on homogenous performance metrics that are time- or distance-based, no consideration is given to the heterogeneity of cases and patients' survival. Erkut et al. (*24*) point out that the trend in EMS response research is to substitute time- and distance-covering concepts with concepts that account for survival probabilities. This type of metric is used in recent works (*9*, *19*, *25*) and is shown to be more suitable when assessing EMS from the user perspective; additionally, it allows for benchmarking comparisons (*24*, *25*).

At the present time, however, vehicle dispatching rules for medical emergency requests follow distance- or time-based metrics such as the classic closest idle vehicle dispatching rule (*12*, *26*, *27*), which consist of allocating the vehicle that is closest to the emergency occurrence site. This rule is tied to the previously mentioned classic performance metrics that focus on the overall system response time. Thus, there is no consideration for the heterogeneity or survivability of emergency cases. Several studies have started to investigate the fact that the closest idle vehicle is not always the optimal solution. The next closest vehicles could still provide acceptable service while the closest vehicle provides better coverage of the network if it remains available in the following hours (*28*, *29*).

In many countries, the EMS does not even consider real-time information when assessing the closest idle vehicle. Instead, they mostly rely on spatial distances or on the operators' skills and experience. Moreover, time-dependent information has been shown to be one of the key factors for better vehicle dispatching rules (*1*).

Clearly, real-time information and different vehicle dispatching rules are the most recent topics in EMS optimization. Travel times and changes with respect to EMS call volumes are used by Schmid (*1*) to achieve a decrease in the average response time of emergency services. Haghani et al. (*30*) propose an optimization model for real-time emergency vehicle dispatching and routing, using real-time traffic information to better support dispatching decisions. Nevertheless, the proposed model presents major issues when applied in real-time situations, owing to its computation time burden. Thus, Haghani and Yang (*12*) upgrade this model by drastically reducing the computational burden using the rolling-horizon approach, and adding coverage concerns for future demand. Undoubtedly, demand prediction presents an interesting challenge for researchers (*19*), as does the use of real-time information to better model dispatching rules. Li et al. (*31*) use the uncertainty theory to deal with uncertain factors, such as demand, when dispatching medical emergency resources. Still, they do not consider environmental randomness, which can be captured using simulation. On the contrary, Knopps and Lundgren (*32*) compile the most common dispatching rules—the classic closest rule, and the preference rule, which tries to minimize the response time for high-priority calls (*33*)—and propose new ones. These are the modified preference rule, which adds to the original a maximum response time threshold for lower-priority calls, and the preparedness rule, which employs a function that measures the preparedness of the system for future calls and tries to maximize it when dispatching vehicles to lower-priority emergencies. In the work of Jagtenberg, et al. (*26*) a benchmark model is proposed and when applied in a case study, the authors were able to show that the closest idle vehicle is a factor of 2.7 away from the optimum.

These state-of-the-art research topics (real-time information, EMS heterogeneity, survival indicators) are scattered and analyzed individually, and thus no relevant research combines them with emergency vehicle dispatching rules performance and assesses their consequences.

Methodological Approach

The literature review clarified that simulation is a superior tool to assess the performance of EMS dispatching rules with realism and accuracy. Furthermore, it was pointed out that although the preferred EMS performance metric is the system response time, survival functions are more suitable for measuring the patients' outcomes.

Consequently, to investigate this work claim, a simulation is proposed using an agent-based model. The authors suggest performance metrics related to response time and survival, and dispatching rules based on the classic rule of closest idle vehicle and survival functions.

The method consists of building a simulation of a real city with real emergency call data to test different scenarios, in which the different dispatching rules are applied in different conditions, such as the existence or not of real-time information and during various periods of the year. The simulation results are calculated based on the chosen performance metrics and a comparison is made.

Simulation Model

An agent-based model is used to test the EMS vehicle dispatching rules and simulate urban environment conditions, in which a city agent controls a group of lower-level agents: events, road network, vehicles, and nodes, as shown in Algorithm 1.

Algorithm 1. Simulation algorithm.

Definitions:
$\boldsymbol{N}$ = set of nodes $\boldsymbol{n}$
$\boldsymbol{n}$ = node, where $\boldsymbol{s}$= node of type station and $\boldsymbol{h}$ = node of type hospital
$\boldsymbol{V}$ = set of vehicles $\boldsymbol{v_s}$
$\boldsymbol{v_s}$ = vehicle in station $\boldsymbol{s}$
$\boldsymbol{S}$ = set of stations $\boldsymbol{s}$
$\boldsymbol{H}$ = set of hospitals $\boldsymbol{h}$
$\boldsymbol{E}$ = set of events $\boldsymbol{e^t_n}$
$\boldsymbol{e^t_n}$ = emergency event occurring at node n during $\boldsymbol{t}$
$\boldsymbol{M}$ = set of matrices $\boldsymbol{M^t}$
$\boldsymbol{M^p}$ = matrix of real travel times for period $\boldsymbol{p}$
$\boldsymbol{T}$ = total simulation time
$\boldsymbol{t}$ = time
step = temporal resolution
f() = programming function
While $\boldsymbol{t < T}$:
1. **Update city( )**"set **t** and activate $\boldsymbol{e^t_n}$"
2. **Update network( )**"interact through every v_s to travel one ***step*** and transfer it to destination nodes"
3. **Update events( )**"activate e^t_n and the vehicle dispatching algorithm"
   – Network calculates time travel from all stations
   – Network returns the shortest time travel
   – Vehicle dispatching algorithm runs
4. **Update vehicles job( )**"updates $\boldsymbol{v_s}$ status"
   – If $\boldsymbol{v_s}$ arrived at $\boldsymbol{e^t_n}$, activate assisting timer
   – If assisting timer ends, request network to be processed to $\boldsymbol{h}$
   – If v_s arrived at $\boldsymbol{h}$, transfers $\boldsymbol{v_s}$ to $\boldsymbol{s}$.
5. **Update results( )**"calculates the EMS performance at the current ***step***"
6. **t = t + *step***

The city agent is the main model agent and is responsible for storing and controlling all other agents by giving them update requests. The EMS calls are simulated by an event agent, which is responsible for activating events and keeping track of their status. When an event is activated, this agent sends an assistance request to the city agent. When assisted by a vehicle agent, the event agent is also responsible for requesting directions to the nearest hospital from the city agent.

To simulate the transport network, a network agent is created with the responsibility of routing all vehicles and assisting in vehicle dispatching and transport. It is also responsible for calculating the fastest origin–destination (OD) routes for different temporal scenarios or network statuses.

A vehicle agent is added to simulate each individual EMS transport/assistance unit. Each vehicle agent keeps track of its position and informs the network agent when it arrives at any destination. This is simulated on a microscale by tracking the distance when traveling between nodes; the delays and assisting times that are randomly generated by the event agent. The vehicle agent transports victims (event agents) from the occurrence node to the hospital node. It is also responsible for keeping track of the station to which it must return after a job is completed.

The node agent can be of three types: network node, hospital node, or station node. This agent assists the network and city agent by storing vehicles and events, and constitutes the origins and destinations of the network agent. The routing is abstracted by a previous calculation of travel times between origins and destinations for different conditions, which are then stored in the network agent. This simplification does not detrimentally affect the methodology, because this research does not focus on vehicle rerouting. Moreover, by simplifying, the simulation model becomes faster, thereby allowing for more complex dispatching algorithms and longer analyzed periods.

Performance Metrics

The performance P_i of an EMS response to an event $\boldsymbol{i}$ of type $\boldsymbol{t}$ can be defined by a function that may depend on the time between the event start and the arrival of an assistance vehicle, $\boldsymbol{r_i}$, given by Equation 1.

$$P_i = f^t(r_i) \tag{1}$$

The classic response time metric, $\boldsymbol{P^c}_i$, can then be described as Equation 2:

$$P_i^c = r_i \tag{2}$$

which is then generalized for the overall service either by its summed parts, Equation 3, or its average, Equation 4:

$$P^c = \sum_{i=1}^{n} r_i \tag{3}$$

$$\overline{P^c} = \frac{\sum_{i=1}^{n} r_i}{n} \tag{4}$$

This metric is sometimes simplified into a binary evaluation, $\boldsymbol{P^T}_i$, Equation 5, where $\boldsymbol{T}$ is a time threshold:

$$\begin{cases} P_i^T = 0 \text{ if } r_i > T \\ P_i^T = 1 \text{ if } r_i \leq T \end{cases} \tag{5}$$

Recently, as pointed out in the literature review, these solely time-based metrics (or distance-based metrics if $\boldsymbol{r_i}$ is substituted by the distance between $\boldsymbol{i}$ and the station that responds to it) can be replaced by a survival function that measures the likelihood of a victim surviving a medical emergency of type $\boldsymbol{e}$ when assisted within $\boldsymbol{r_i}$, in line with the work of Erkut et al. (*24*), Equation 6:

$$P_i^s = \left(e^{c^e \times r_i}\right)^{-1} \tag{6}$$

where $\boldsymbol{c^e}$ is a survival coefficient for an event of type $\boldsymbol{e}$. This metric can also be generalized to the system by taking the form of a sum or an average, Equation 7:

$$P^s = \sum_{i=1}^{n} \left(e^{c^e \times r_i}\right)^{-1} \text{ or } \overline{P^s} = \frac{\sum_{i=1}^{n} \left(e^{c^e \times r_i}\right)^{-1}}{n} \tag{7}$$

Dispatching Rules

Three dispatching rules are proposed for this investigation, as shown in Algorithm 2. The classic rule, in which the closest vehicle is dispatched, is named ClosestDR. When this rule uses real-time information of traffic it becomes RT-ClosestDR. A random vehicle dispatching rule, RT-RandomDR, is considered to determine how choosing the closest vehicle might compare with choosing any other vehicle, proving the inadequacy of the classic rule.

Finally, an intelligent survival dispatching rule, RT-IntelligentSurvivalDR, is proposed here. The survival performance metric is used to calculate the system status after dispatching a certain vehicle, and the intelligent algorithm maximizes the system status, $\boldsymbol{S^p}$, at period $\boldsymbol{s}$ by dispatching to non-life-threatening emergencies the vehicle that penalizes $\boldsymbol{S^p}$ the least. This is calculated using the busy fraction $\boldsymbol{q_s}$ (*16*) and the expected response time by Berg et al. (*34*), according to Snyder and Daskin (*35*):

Algorithm 2. Dispatching rules.

While no $\mathbf{v_s}$ is dispatched:
If ClosestDR or RT-ClosestDR is active:
1. Sort $\boldsymbol{S}$ by ascending $\boldsymbol{r_i}$
2. From index 0 to size of **S**, dispatch $\mathbf{v_s}$ if idle

If RT-RandomDR is active:
3. Select all s within $\boldsymbol{r_i} <$ maximum response limit which have at least one idle $\mathbf{v_s}$. If none, go to step 5.
4. Dispatch a random v_s from step 3.
5. Go to step 1.

If RT-IntelligentSurvivalDR:
6. Select all s within $\boldsymbol{r_i} <$ maximum response limit which have at least one idle v_s. If none, go to step 10.
7. For every s in step 6, calculate system status using Equation 11.
8. Sort solutions from step 7 by descending $\mathbf{Q_s}$
9. From index 0 to size of solution from step 7, dispatch $\mathbf{v_s}$ if idle.
10. Go to step 1

$$S^p = \sum_i \sum_s \sum_k \sum_e d_i^{es} \left(e^{c^e \times r_{is}^p}\right)^{-1} \times (1 - q_s) \times q^{k-1} \times z_{kis} \tag{8}$$

where:

$$q = \frac{\sum_e d_i^{ep}}{a_s} \tag{9}$$

$$q^{k-1} = \prod_{j=1}^{k-1} q_j \tag{10}$$

$\boldsymbol{d^{es}}_i$ is the predicted number of events of type $\boldsymbol{e}$ during period $\boldsymbol{s}$ at demand point $\boldsymbol{i},\boldsymbol{a_j}$ is the number of idle vehicles allocated at station $\boldsymbol{s}$, and z_{kis} is 1 if $\boldsymbol{s}$ is the $\boldsymbol{k^{th}}$ nearest station to $\boldsymbol{i}$, 0 otherwise.

Thus, the decision to dispatch a vehicle from a station $\boldsymbol{s}$ to a certain event can be scored as $\boldsymbol{Q_s}$:

$$Q_s = \left(e^{c^e \times r_{is}^p}\right)^{-1} + S^{p'} \tag{11}$$

where $\boldsymbol{S^{p'}}$ is the system survival status when an ambulance from $\boldsymbol{s}$ was dispatched to $\boldsymbol{i}$, updating $\boldsymbol{a_s}$ to $\boldsymbol{a_s}-1$.

Both the RT-RandomDR and RT-Intelligent SurvivalDR algorithms can be controlled by a threshold time to ensure that any emergency is assisted within a maximum response time limit. If no vehicle within this maximum response time limit is available, then the closest vehicle is dispatched. Additionally, they are only activated for non-life-threatening emergencies.

Application of the Model

To assess the validity of this work's arguments and the performance of the different dispatching rules, the methodology is applied to a simulation created with real data from San Francisco, Figure 1.

The data was collected from the U.S. Government's open data strategic American resources and consists of a collection of Calls-For-Service database, which includes all the fire units' responses to calls in a total of 4.4 million vehicles dispatched between 2000 and 2017. The fire department is responsible for managing the EMS calls and responses, requesting a private unit when required. Thus, the dispatch of private units (to 911 calls) is also recorded in the database.

The data were processed and filtered into a SQL database for easy access and data manipulation. From the same open data source, fire station locations were acquired and added to the SQL database.

The city was divided in a grid of 500 m × 500 m cells (Figure 1*b*) and each unit is represented by a node corresponding to its center, with a total of 518 nodes. The fire stations, a total of 45, were then assigned to the closest node as well as three fictional hospitals representing the major San Francisco hospitals. The lack of information on the destination hospital for each call necessitated a random allocation of calls to hospitals based on their proximity. It would be possible to identify the hospital most likely to receive the victims using the San Francisco EMS assignment algorithm. However, the required information for this allocation is also not present in the database.

To implement the system performance metrics and the simulation model, travel time matrices were built using real travel times collected from Google through its Directions API. This specific API allows the calculation of travel times for different days and hours. Because of the burden of this process and the API limitations, a resolution of week versus weekend with 3-h intervals was used when building the OD matrices.

The intelligent vehicle dispatching algorithm coefficients were trained with calls from 2010 to 2015 and the priority calls were defined as those tagged "Potentially Life-Threatening" in the original data. Finally, several runs of the simulation model were computed for different months of the year of 2016, different numbers of vehicles, and the different dispatching rules.

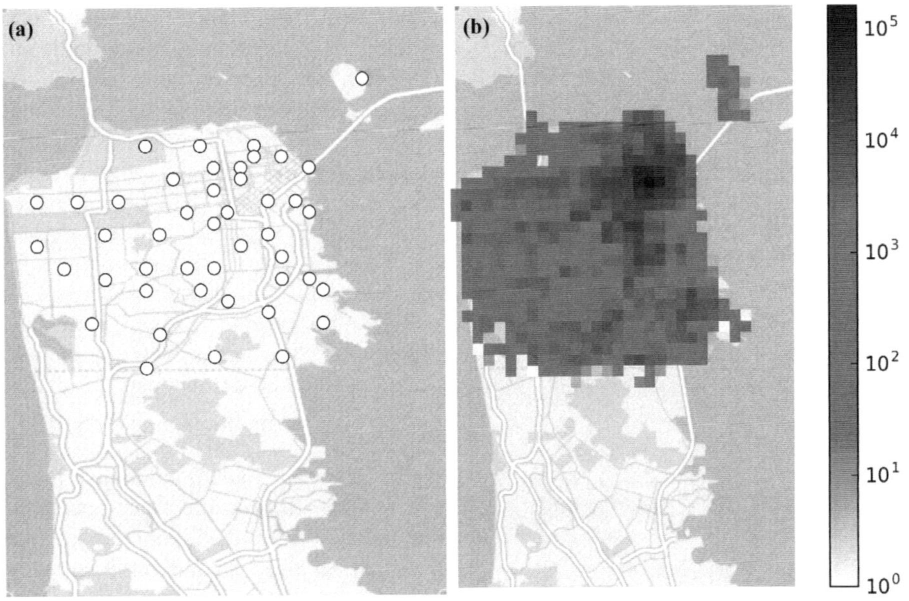

Figure 1. San Francisco experiment area with (*a*) the available stations and (*b*) the total number of calls (2000–2017). Each circle in (*a*) represents a station, and each square in (*b*) represents one grid cell.

All the scripts for collecting the data and for the simulation model were programmed in Python 2.7 and were run on a machine with an Intel core i7 quad core processor at 1.73 GHz and 8 GB of RAM running a Windows 10 64-bit operating system. The running time for each experiment depends on the dispatching algorithm used and the number of simulated days, but each experiment took between 20 s and 4 h.

The relevant results were processed and compiled in comparative graphs, which are analyzed in the next chapter.

Results

To analyze the performance of each dispatching rule for "Potentially Life-Threatening" emergency events, the ClosestDR algorithm is set as the base rule for comparison. Each rule is then tested for different scenarios, and different metrics are calculated: the survival rate, average response time, and number of victims assisted within 8 min. The 8-min threshold is used because this is a common threshold used in several works (*9, 19*). For a better understanding of the results, several graphs were compiled, in which the accumulative performance of each dispatching rule is calculated as a gain over the base rule.

The methodology was first applied to the month of February and tested for each dispatching rule in configurations of one vehicle per station and two vehicles per station. This allows for an understanding of how the number of available vehicles might influence the EMS response and how each dispatching rule performs in each condition.

Table 1 presents a summary of the different performance metrics for the different configurations and dispatching rules, and Figure 2 shows the gain of each dispatching rule when compared with the base rule. There is an obvious increase in performance when the number of vehicles doubles. One of the reasons for this is the fact that, with the existence of two vehicles per station, there are no delays in responding to emergency calls. This means that at least one vehicle is always idle and available to be dispatched.

Interestingly, the random dispatching rule, RT-RandomDR, outperforms all other rules when resources are scarce (one vehicle per station). Nevertheless, when measuring the performance by the number of victims assisted within 8 min, the RT-IntelligentSurvivalDR is shown to be optimal, and as expected, the RT-ClosestDR algorithm provides the overall lowest average response time when all calls are considered.

From Figure 2, a break is visible in the tendency of each rule's performance around the 4000th event. Further research revealed that this disruption corresponds to the period when the "Superbowl 50" event was held near San Francisco, which led to the occurrence of several smaller events across the city, and probably an increase in the number of tourists/visitors. This also shows that in such an unpredictable situation, the

Table 1. Summary of the Results for the Month of February with Two Different Configurations: One Vehicle per Station and Two Vehicles per Station and Different Performance Indicators

Dispatching rule	All calls				Possible life-threatening					
	Average response (s)		Average delay (s)		Average response (s)		Total events assisted in less than 8 min		Average survival rate per event	
No. vehicles/station	1	2	1	2	1	2	1	2	1	2
ClosestDR	585	337	24	0	634	348	5116	9463	42%	59%
RT-ClosestDR	520	287	64	0	601	295	6086	10397	45%	64%
RT-RandomDR	563	396	26	0	536	290	6617	10526	48%	64%
RT-IntelligentSurvivalDR	609	329	36	0	558	288	6447	10571	47%	64%

ClosestDR algorithm performs better than the nonrandom rules, as it minimizes the delays experienced during demand peaks. This can be ascribed to the fact that this algorithm acts as a restricted random dispatching rule, because it does not use real-time information.

To further investigate the problem of sporadic events that may overload the EMS response and produce system delays, a simulation was placed for the period around New Year's Eve (from December 30, 2016 to January 3, 2017), which is believed to be an event likely to generate a high demand peak and to which readers may relate.

The results, shown in Figure 3, indicate that when the EMS resources (number of vehicles) are low, the system performance degrades independently of the rule applied, which is explained by the existence of delays. In a situation where resources are not scarce, each dispatching rule seems to perform similarly, recovering its usual performance when the demand peak vanishes. Moreover, it is important to note that the time response metric is the most sensitive one in encumbered periods and under low resources, because it is the only metric that captures delays.

Finally, the seasonal adaptability of each dispatching rule performance is analyzed, as shown in Figure 4, by running simulations of different months. For the simulations where the number of vehicles is scarce, the intelligent survival algorithm underperforms in nearly all of the analyzed months, owing to delay propagation. The exception is February, in which the closest dispatching rule is the weaker rule. In the other months, the best performance is achieved either by the closest rule or the random rule. When the number of vehicles per station is doubled, the intelligent survival dispatching rule again becomes the best practice, although the random dispatching rule performs in a similar fashion.

Apparently, the lack of sufficient resources to respond in a timely manner to every life-threatening emergency significantly affects the system's response by creating delays. This becomes more obvious for the RT-IntelligentSurvivalDR algorithm and when the performance is measured by the accumulated response time. This gap is less noticeable in the other performance metrics.

Conclusions

This paper presents a methodology to assess the performance of EMS dispatching rules, providing several performance metrics and dispatching rules, including a dispatching rule proposed by the authors that aims to increase victims' survival, without excessively penalizing non-life-threatening emergencies. Furthermore, the method is simulated with real data from the city of San Francisco.

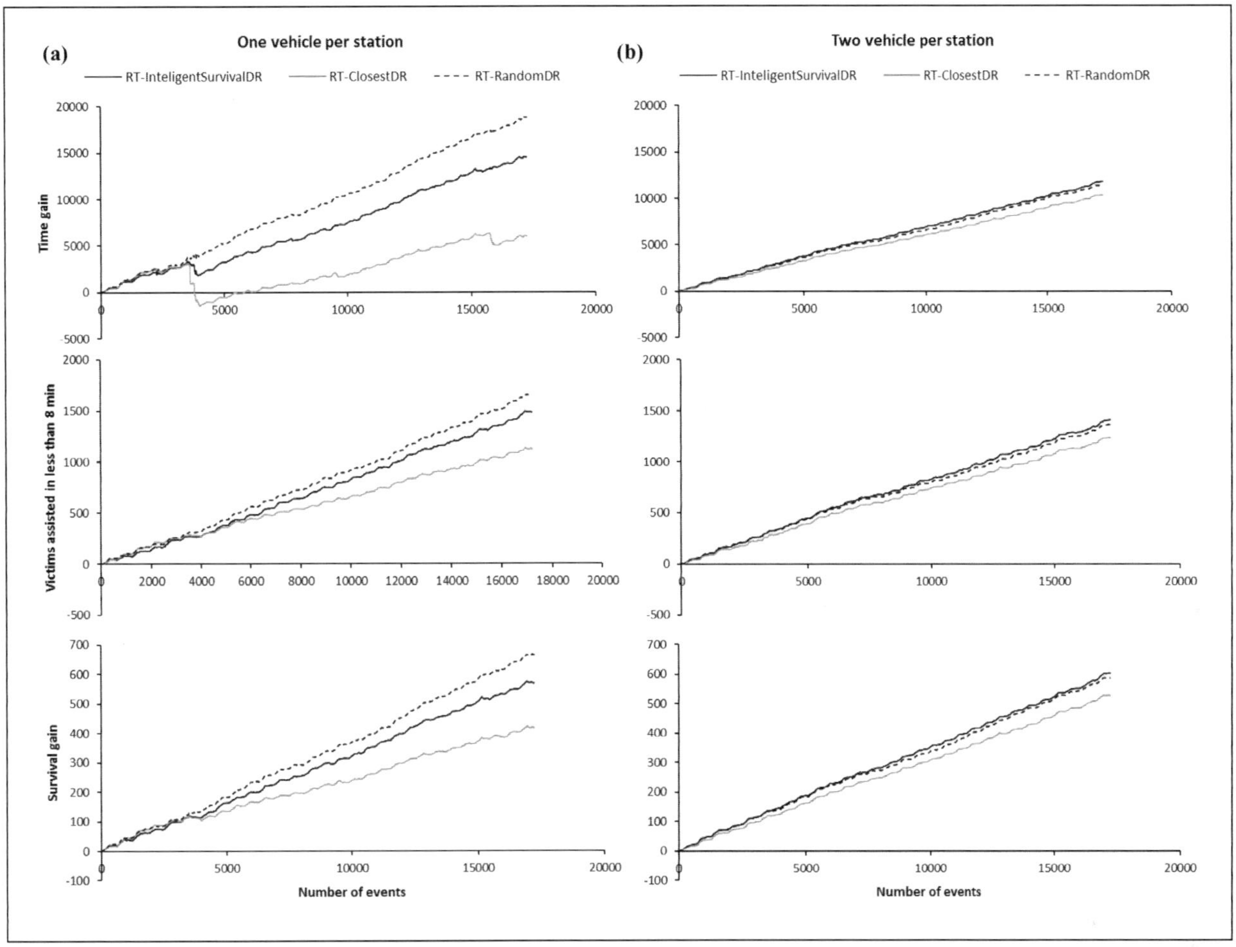

Figure 2. Gain of each dispatching rule when compared with the base rule, ClosestDR. Performance of dispatching rules for a configuration of (*a*) one vehicle per station and (*b*) two vehicles per station.

The presented results open new pathways to the study of EMS vehicle dispatching rules, their consequences, and the characteristics that influence their performance. This work makes an important contribution to the state of the art by providing a proof of concept and empirical results, which illustrate that the optimal dispatching rule is not the obvious one—sending the closest vehicle. They also show that having a single rule is never the optimum in the long term.

In an EMS system in which resources are not scarce, the proposed intelligent survival dispatching rule is the obvious choice in every scenario. Moreover, a random dispatch of vehicles (without penalizing non-life-threatening events by more than 15 min) can outperform the classic rule in almost every tested case.

This work proves that the use of real-time traffic information is essential in every situation, and significantly improves the system's performance in terms of response time, survival rate, and the total number of emergencies assisted within 8 min.

It was found that demand peaks, owing to special occasions on which the number of people in the same area increases or when people engage in more risky behaviors, present an interesting challenge, and the optimal solution for such cases is difficult to determine with the presented dispatching rules. It came to the authors' attention that the lower performance of the dispatching rules in this situation is due to the poor handling of delays. It is also possible that different dispatching rules may fit better to different configurations of demand: local demand peaks, or dispersed demand peaks. This is a topic worth investigating.

It is important to note that the emergency patients in a waiting list for a vehicle were processed using a greedy algorithm that tries to minimize the total number of delays by avoiding delay propagation. This is clearly not the optimal approach, and thus the authors propose the creation of delay-handling algorithms that can consider the same EMS performance principles that were discussed through this paper.

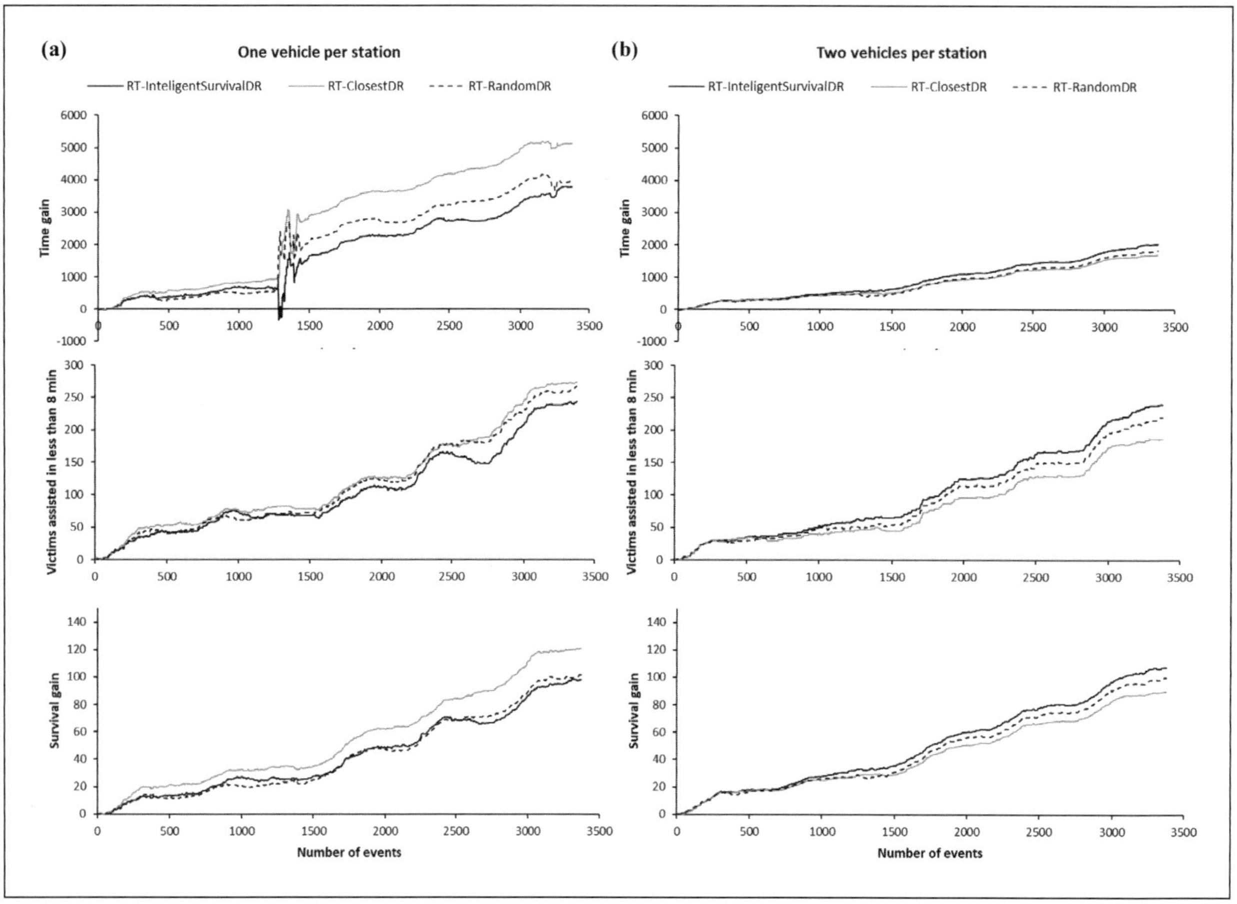

Figure 3. Microanalysis of the gain in performance of each of the dispatching rules for the New Year's Eve event: (*a*) one vehicle per station; (*b*) two vehicles per station.

The simulation model accounts for "in situ" treatment time. Nevertheless, the survival function studies found on the state of the art do not account for hospital care and how pre-hospital time (which includes transport to a hospital) might further affect victims' outcomes (*36–38*). Rapid diagnoses and hospital interventions are key factors affecting victims' survival. This is an interesting topic of investigation and could allow EMS planners to manage resources better, that is, the type and quantity of vehicles to dispatch to each event, allowing the phased dispatch of treatment and transport units, and reallocation of the treatment unit to another event after "in situ" assistance.

The presented topic could be further advanced with the use of additional details in the simulation model and response policies, such as the use of different vehicle types, vehicle rerouting, and a wider discretization of heterogeneity of emergency cases. Moreover, with the application of these concepts to other cities to further support this study's conclusions, practitioners will have more tools to implement better dispatching policies and/or identify critical problems in their response systems and make better tactical decisions.

This work highlights the fact that sending the closest vehicle (for non-life-threatening emergencies) to a certain demand point tends to debilitate that same demand point and its surroundings. This is because, statistically speaking, the same demand point proves to have a higher chance of having more emergencies, and thus the obvious dispatching rule is not always the optimal one.

Acknowledgments

We acknowledge the support of FCT (Portuguese national funding agency for science, research and technology) under the grant PD/BD/52355/2013 during the development of this work.

Author Contributions

The authors confirm contribution to the paper as follows: study conception and design: Marco Amorim and António Couto; data collection: Marco Amorim and Sara Ferreira; analysis and interpretation of results: Marco Amorim, António Couto and

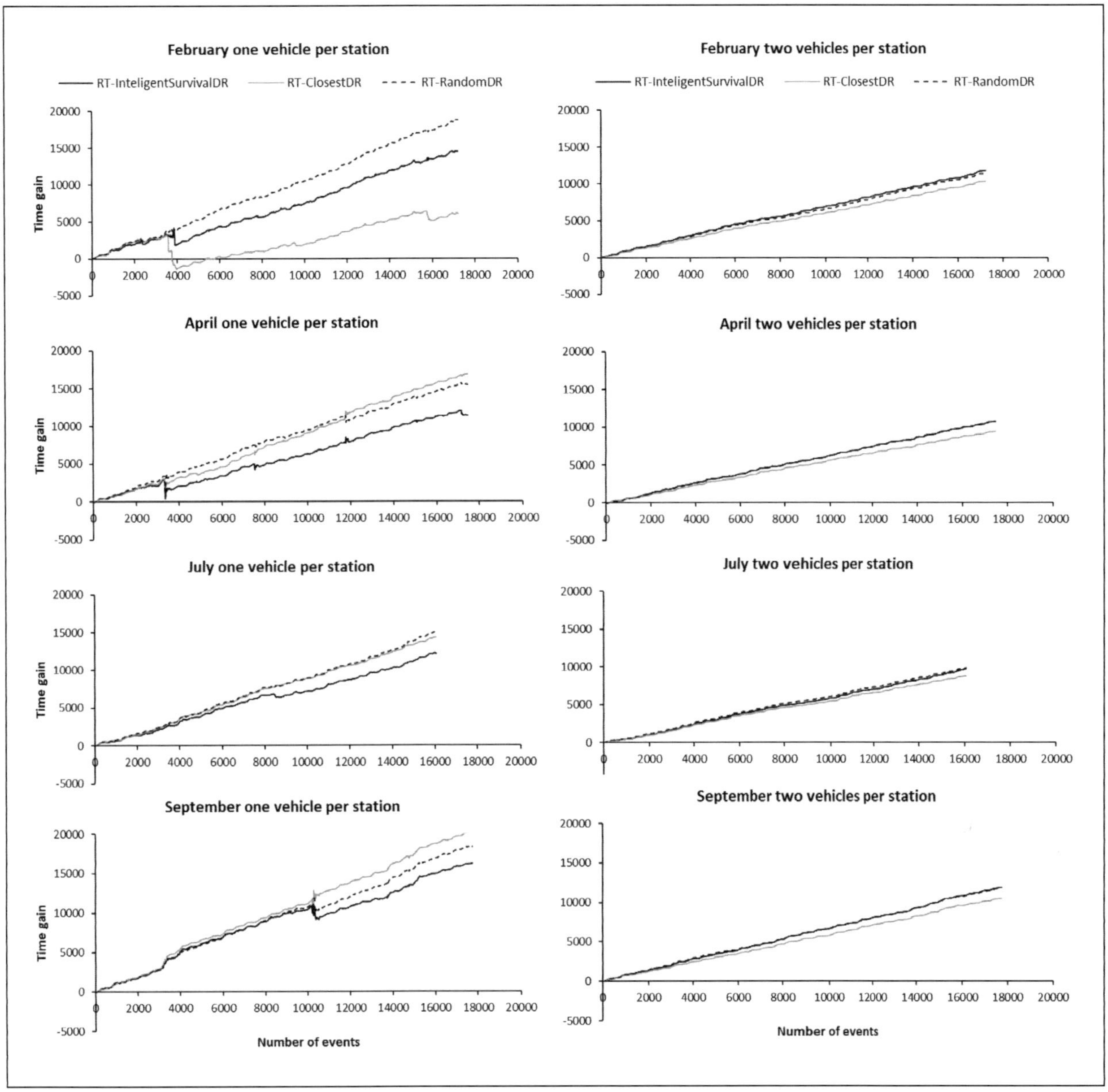

Figure 4. Comparison of the gain in response time performance when compared with the base rule for different months of the year and different vehicle configurations.

Sara Ferreira; draft manuscript preparation: Marco Amorim. All authors reviewed the results and approved the final version of the manuscript.

References

1. Schmid, V. Solving the Dynamic Ambulance Relocation and Dispatching Problem Using Approximate Dynamic Programming. *European Journal of Operational Research*, Vol. 219, 2012, pp. 611–621.
2. Ferrucci, F, and S. Bock. Real-time Control of Express Pickup and Delivery Processes in a Dynamic Environment. *Transportation Research Part B: Methodological*, Vol. 63, 2014, pp. 1–14.
3. Valinsky, D. Symposium on Applications of Operations Research to Urban Services—A Determination of the Optimum Location of Fire-Fighting Units in New York City. *Journal of the Operations Research Society of America*, Vol. 3, 1955, pp. 494–512.

4. Hogg, J. M. The Siting of Fire Stations. *Journal of the Operations Research Society of America*, Vol. 19, 1968, pp. 275–287.
5. Savas, E. S. Simulation and Cost-Effectiveness Analysis of New York's Emergency Ambulance Service. *Management Science*, Vol. 15, 1969, pp. B-608–B-627.
6. Restrepo, M., S. G. Henderson, and H. Topaloglu. Erlang Loss Models for the Static Deployment of Ambulances. *Health Care Management Science*, Vol. 12, 2008, pp. 67–79.
7. Maxwell, M. S., M. Restrepo, S. G. Henderson, and H. Topaloglu. Approximate Dynamic Programming for Ambulance Redeployment. *INFORMS Journal on Computing*, Vol. 22, 2010, pp. 266–281.
8. Yue, Y., L. Marla, and R. Krishnan. An Efficient Simulation-Based Approach to Ambulance Fleet Allocation and Dynamic Redeployment. *Association for the Advancement of Artificial Intelligence*, Toronto, Canada, July 2012. Available at: https://www.aaai.org/ocs/index.php/AAAI/AAAI 12/paper/view/5148 Accessed June 1, 2018.
9. McCormack, R., and G. Coates. A Simulation Model to Enable the Optimization of Ambulance Fleet Allocation and Base Station Location for Increased Patient Survival. *European Journal of Operational Research*, Vol. 247, 2015, pp. 294–309.
10. Iannoni, A. P., R. Morabito, and C. Saydam. An Optimization Approach for Ambulance Location and the Districting of the Response Segments on Highways. *European Journal of Operational Research*, Vol. 195, 2009, pp. 528–542.
11. Su, S., and C.-L. Shih. Modeling an Emergency Medical Services System Using Computer Simulation. *International Journal of Medical Informatics*, Vol. 72, 2003, pp. 57–72.
12. Haghani, A., and S. Yang. Real-Time Emergency Response Fleet Deployment: Concepts, Systems, Simulation & Case Studies. In: *Dynamic Fleet Management: Concepts, Systems, Algorithms & Case Studies* (V. Zeimpekis, C. D. Tarantilis, G. M. Giaglis and I. Minis eds.), Springer, Boston, MA, 2007, pp. 133–162.
13. Toregas, C., R. Swain, C. ReVelle, and L. Bergman. The Location of Emergency Service Facilities. *Operations Research*, Vol. 19, 1971, pp. 1363–1373.
14. Church, R., and C. R. Velle. The Maximal Covering Location Problem. *Papers in Regional Science*, Vol. 32, 1974, pp. 101–118.
15. Daskin, M. S., and E. H. Stern. A Hierarchical Objective Set Covering Model for Emergency Medical Service Vehicle Deployment. *Transportation Science*, Vol. 15, 1981, pp. 137–152.
16. Daskin, M. S. A Maximum Expected Covering Location Model: Formulation, Properties and Heuristic Solution. *Transportation Science*, Vol. 17, 1983, pp. 48–70.
17. Hogan, K., and C. ReVelle. Concepts and Applications of Backup Coverage. *Management Science*, Vol. 32, 1986, pp. 1434–1444.
18. ReVelle, C., and K. Hogan. The Maximum Availability Location Problem. *Transportation Science*, Vol. 23, 1989, pp. 192–200.
19. Amorim, M., S. Ferreira, and A. Couto. Road Safety and the Urban Emergency Medical Service (uEMS): Strategy Station Location. *Journal of Transport & Health*, Vol. 6, 2017, pp. 60–72.
20. Blackwell, T. H., and J. S. Kaufman. Response Time Effectiveness: Comparison of Response Time and Survival in an Urban Emergency Medical Services System. *Academic Emergency Medicine*, Vol. 9, 2002, pp. 288–295.
21. Pons, P. T., J. S. Haukoos, W. Bludworth, T. Cribley, K. A. Pons, and V. J. Markovchick. Paramedic Response Time: Does It Affect Patient Survival? *Academic Emergency Medicine*, Vol. 12, 2005, pp. 594–600.
22. Sánchez-Mangas, R., A. García-Ferrrer, A. de Juan, and A. M. Arroyo. The Probability of Death in Road Traffic Accidents. How Important Is a Quick Medical Response? *Accident Analysis & Prevention*, Vol. 42, 2010, pp. 1048–1056.
23. Valenzuela, T. D., D. J. Roe, G. Nichol, L. L. Clark, D. W. Spaite, and R. G. Hardman. Outcomes of Rapid Defibrillation by Security Officers after Cardiac Arrest in Casinos. *New England Journal of Medicine*, Vol. 343, 2000, pp. 1206–1209.
24. Erkut, E., A. Ingolfsson, and G. Erdogan. Ambulance Location for Maximum Survival. *Naval Research Logistics*, Vol. 55, 2008, pp. 42–58.
25. Knight, V. A., P. R. Harper, and L. Smith. Ambulance Allocation for Maximal Survival with Heterogeneous Outcome Measures. *Omega*, Vol. 40, 2012, pp. 918–926.
26. Jagtenberg, C. J., P. L. van den Berg, and R. D. van der Mei. Benchmarking Online Dispatch Algorithms for Emergency Medical Services. *European Journal of Operational Research*, Vol. 258, 2017, pp. 715–725.
27. Yang, S., M. Hamedi, and A. Haghani. Online Dispatching and Routing Model for Emergency Vehicles with Area Coverage Constraints. *Transportation Research Record: Journal of the Transportation Research Board*, 2005. 1923: 1–8.
28. Jagtenberg, C. J., S. Bhulai, and R. D. van der Mei. Dynamic ambulance dispatching: is the closest-idle policy always optimal? *Health Care Management Science*, Vol. 20, No. 4, 2017, pp. 517–531. https://doi.org/10.1007/s10729-016-9368-0
29. Carter, G. M., J. M. Chaiken, and E. Ignall. Response Areas for Two Emergency Units. *Operations Research*, Vol. 20, 1972, pp. 571–594.
30. Haghani, A., H. Hu, and Q. Tian. *An Optimization Model for Real-time Emergency Vehicle Dispatching and Routing*, 2003.
31. Li, H., J. Peng, S. Li, and C. Su. Dispatching Medical Supplies in Emergency Events via Uncertain Programming. *Journal of Intelligent Manufacturing*, Vol. 28, 2017, pp. 549–558.
32. Knopps, L., and T. Lundgren. *Modeling Ambulance Dispatching Rules for EMS-Systems*, 2016. Available at: http://urn.kb.se/resolve?urn=urn:nbn:se:kth:diva-189023.
33. Bandara, D., M. E. Mayorga, and L. A. McLay. Priority Dispatching Strategies for EMS Systems. *Journal of the Operational Research Society*, Vol. 65, 2014, pp. 572–587.
34. van den Berg, P. L., J. Theresia van Essen, and E. J. Harderwijk. Comparison of static ambulance location models.

Proc., 3rd International Conference on Logistics Operations Management (GOL), IEEE, 2016, pp. 1–10.
35. Snyder, L. V., and M. S. Daskin. Reliability Models for Facility Location: The Expected Failure Cost Case. *Transportation Science*, Vol. 39, 2005, pp. 400–416.
36. Kereiakes, D. J., W. D. Weaver, J. L. Anderson, T. Feldman, B. Gibler, T. Aufderheide, D. O. Williams, L. H. Martin, L. C. Anderson, J. S. Martin, G. McKendall, M. Sherrid, H. Greenberg, and S. L. Teichman. Time Delays in the Diagnosis and Treatment of Acute Myocardial Infarction: A Tale of Eight Cities. Report from the Pre-hospital Study Group and the Cincinnati Heart Project. *American Heart Journal*, Vol. 120, 1990, pp. 773–780.
37. Curtis, J. P., E. L. Portnay, Y. Wang, R. L. McNamara, J. Herrin, E. H. Bradley, D. J. Magid, M. E. Blaney, J. G. Canto, and H. M. Krumholz. The Pre-Hospital Electrocardiogram and Time to Reperfusion in Patients with Acute Myocardial Infarction, 2000–2002: Findings from the National Registry of Myocardial Infarction-4. *Journal of the American College of Cardiology*, Vol. 47, 2006, pp. 1544–1552.
38. Gaspoz, J. M., P. F. Unger, P. Urban, J. C. Chevrolet, W. Rutishauser, C. Lovis, L. Goldman, C. Héliot, L. Séchaud, S. Mischler, and F. A. Waldvogel. Impact of a Public Campaign on Pre-hospital Delay in Patients Reporting Chest Pain. *Heart*, Vol. 76, 1996, pp. 150–155.

The Standing Committee on Transportation Safety Management (ANB10) peer-reviewed this paper (18-00049).

Article

TRR
JOURNAL OF THE TRANSPORTATION RESEARCH BOARD

Transportation Research Record
2018, Vol. 2672(32) 22–29

DOI: 10.1177/0361198118786814
journals.sagepub.com/home/trr
SAGE

Student Pedestrian Walking Speeds at Crosswalks Near Schools

Kevin Chang[1], Peter Foss[2], Meagan Larrea[1], and Edinson Bautista[1]

Abstract

The percentage of young people walking to school has decreased significantly since the 1960s. The reasons for this decline have been attributed to working parents too busy to accompany their children on a daily walk to school, parental concerns about safety from "stranger danger," and concerns about the risks of walking routes that must cross at least one higher speed or higher volume roadway. School crossings at arterial facilities are often necessitated by the fact that newer schools are sited in peripheral locations and are not located within the heart of residential communities where sidewalks and slower traffic are expected. The objective of this study was to measure and assess the walking speeds of today's young people who attend either an elementary, middle, or high school and to determine how their walking behavior compares with existing guidelines. The walking speed parameter is a critical component that is used to determine the duration of flashing beacons, high-intensity activated crosswalk (HAWK) beacons, and other pedestrian-activated devices; these devices facilitate school walking routes and provide extra assurance for children and parents alike. This study concluded that the walking speeds of school-aged children, even at the 15th percentile, are generally higher than those of current guidelines, suggesting that agency practitioners have an opportunity to fine tune timing parameters to reduce delay for the motoring public while still ensuring an appropriate and necessary level of safety for school-aged children.

In 1969 in the United States nearly half of all children walked or bicycled to school, rising to 89% of those living within one mile of their destination. By 2009, only 13% walked or bicycled, and only 35% of those living within one mile of school continued to do so (*1*). Recent studies have shown a slight increase in walking percentages, to 15% in 2013, due to increased awareness and educational campaigns (*2*). This multi-generational reduction is anecdotally attributed to a number of societal factors including: dual-income parents who have less time to accompany their son or daughter to and from school; reluctance of parents to allow their children to walk or bicycle to school due to concerns with regard to child safety and security; and school siting decisions, based on the cost and availability of land, that make walking and bicycling less feasible due to the remote locations of newer schools. As an overall result, the school transportation-related ramifications include increased traffic near schools and missed opportunities for physical fitness by children at a time of increased childhood obesity.

The potential long-term effects of this reduction in walking are concerning. In the state of Idaho, obesity rates for the general population have tripled from 10% in 1990 to nearly 30% in 2013 (*3*). A fundamental shift in this trend requires both healthy living and good habit-forming behavior by the next generation of young people, and embracing non-motorized transport in the form of walking (or bicycling) to school from kindergarten to grade 12 (K–12) is an important first step. Pedestrian travel has become an important aspect of transportation as individuals seek mobility opportunities and develop habits to lead a healthier lifestyle. Pedestrian travel is typically confined to sidewalks and multiuse paths, as usage on these facilities tends to be safe and easier to design. The complexity of pedestrian travel increases at intersections when there are interactions with vehicles. It is understood that some of the most dangerous areas for vehicles occur at intersections where multiple conflict points potentially exist; this is also true for pedestrians whether the crossing is marked or unmarked.

Understanding how pedestrians use crosswalks is vital to locating, designing, and establishing crosswalks that are both safe and effective. For a school-aged child, a crosswalk located on a busier roadway that is perceived to be unsafe by his or her parents may represent the single determining factor as to whether or not that child will be allowed to walk (or bicycle) to school. For this reason, a deeper understanding of how school-aged children are using crosswalks is needed and one of the primary characteristics is the speed of each pedestrian;

[1]University of Idaho, Moscow, ID
[2]Transportation Engineer, HDR, Boise, ID

Corresponding Author:
Address correspondence to Kevin Chang: kchang@uidaho.edu

Table 1. Selected Walking Speed Studies

Author(s)	Student age range (years old)	Sample size	Average walking speed (fps)	Year
Sutherland, et al. (*6*)	1 to 7	169	3.7	1988
Knutson, Schimmel, and Ruff (*4*)	6 to 13	227	4.3–5.5	1999
David and Sullivan (*5*)	5 to 12	370	3.7–4.7	2005
Christina de David, et al. (*7*)	3 to 6	20	3.7–4.3	2006
Stansfield, et al. (*8*)	7 to 12	16	2.4–5.4	2006
Dusing and Thorpe (*9*)	1 to 10	438	2.7–4.4	2007
Lythgo, Wilson, and Galea (*10*)	5 to 27*	898	4.3–4.6	2009

*Study included young adults.

this speed will determine the appropriate duration of any actuated or pedestrian-activated device.

Literature Review

A number of studies have examined walking behavior and the personal attributes (such as age and gender) affecting actual walking speeds, with some focused on school-aged children. Walking speed studies have been conducted in a variety of different formats in different settings, with observational data of walking speeds near elementary schools representing one such method. In a 1999 Iowa study, researchers sought to establish a walking speed database based on a 30-second time interval. A total of 227 children under the age of thirteen were recruited from public schools throughout the state and average speeds of 4.3 feet per second (fps), 4.7 fps, 5.5 fps, 5.4 fps, 5.5 fps, 5.5 fps, and 5.0 fps were recorded for students aged six to twelve years, respectively (*4*). Walking speed in hallways was collected in another elementary school mobility study in which students walked in line as a class behind a chosen line leader in a hallway. The amount of time required by class-lines to walk 50 feet decreased as students progressed in age from kindergarten to the sixth grade. Average speeds of 3.7 fps, 3.8 fps, 4.0 fps, 4.1 fps, 4.3 fps, 4.6 fps, and 4.7 fps were recorded for each respective grade level, beginning with kindergarten children and ending with sixth graders (*5*).

To measure walking speeds, gait studies have been previously conducted. Average walking speeds of 3.7 fps for children were collected for six- and seven-year-old children in one study (*6*). When speed and stride length data during normal walking of three- to six-year-old children were measured amongst 20 healthy children (10 boys and 10 girls), average speeds of 1.14 to 1.15 meters per second (3.7 to 3.8 fps) were recorded for five year olds and 1.29 to 1.32 meters per second (4.2 to 4.3 fps) for six year olds (*7*). A separate study examined the annual gait patterns of 16 children between the ages of seven and twelve years (8 boys and 8 girls) over five consecutive years, and the speed of the children ranged from 2.4 to 5.4 fps, with seven children walking at a rate from 2.4 to 3.0 fps, 70 from 3.0 to 3.6 fps, 208 from 3.6 to 4.2 fps, 142 from 4.2 to 4.8 fps, and 30 in the normalized speed band from 4.8 to 5.4 fps; the children walked at their self-selected speed at all times (*8*). Another study assessed a total of 438 children between the ages of one and ten years who completed two walks of at least three steps each at a self-selected speed. The mean self-selected velocity ranged from an average of 2.7 fps to 4.4 fps with the largest variability occurring in subjects under seven years of age (*9*). Approximately 900 primary school-aged children between the ages of five and thirteen years completed six to eight walks in a different speed study. Outcome measures of speed and stride length increased with age while measured cadence was reduced. The gait speed for five to twelve year olds was shown to be 4.3 fps, 4.4 fps, 4.4 fps, 4.6 fps, 4.4 fps, 4.6 fps, 4.6 fps, and 4.6 fps, respectively (*10*). Table 1 summarizes the broad range of walking speeds that have been determined in the previous studies just described.

Comparable studies focusing on general populations have yielded walking speed results in a similar range. One study examined individuals between the ages of 14 and 64 years and found that the average walking speed was 4.0 fps for all adults but decreased to 3.2 fps for pedestrians over the age of 65 (*11*). Age has traditionally been associated with having the most significant effect on walking speed. One study that collected 1,947 observations determined an average adult walking speed of 4.6 fps, but the average child assisted by an adult had a mean walking speed of 4.0 fps. Based on the data collected, this study recommended a 3.8 fps walking speed for timing pedestrian clearance intervals at locations such as downtown areas, shopping areas, neighborhoods, and school areas, and a 4.0 fps walking speed only for locations with very few older pedestrians, assisted children, or disabled persons, such as college campuses (*12*).

Motorist versus Pedestrian Behavior

Understanding how pedestrians use crosswalks and how drivers react to them is vital to designing crossings that are both safe and effective to the flow of traffic. Traffic safety researchers have long contended that driver behavior outweighs physical elements (such as road design) as a causal factor in motor vehicle collisions (*13*), and when a pedestrian–vehicle collision occurs the behavior of both the driver and pedestrian must be examined.

It has been debated over the years whether marked or unmarked crossings provide a safer environment for people to cross. A 2001 study looked at the effects of crosswalk markings on driver and pedestrian behavior at unsignalized intersections by conducting a before-and-after evaluation of crosswalk markings at eleven locations in four American cities. To test driver reaction, driver speeds were measured at a crosswalk that was initially unmarked and then marked. Speeds were collected when no pedestrian was present (base condition), when a pedestrian was present standing at the edge of the road looking to cross (pedestrian looking), and when the pedestrian had taken one step into the lane of travel as if they were going to cross (pedestrian stepping). It was found that there was no statistical difference in speeds for the base condition since drivers were not required to slow down if a pedestrian was not present. However, there was a significant difference in speed for the pedestrian looking and pedestrian stepping conditions, confirming that motorists respond differently to crosswalks when they were marked. This decrease in speed was believed to indicate an increased driver awareness of pedestrians (*14*). A study by Knoblauch and Raymond (*15*) identified and recorded driver reaction to pedestrian behavior at a crosswalk. The looking behavior of pedestrians at marked and unmarked crosswalks was staged to determine if a driver would respond differently depending on whether or not a pedestrian was actively seeking to cross. The study found that drivers tended to slow down more for pedestrians who appeared to be paying less attention than for those who were actively looking at oncoming traffic.

A British study investigated whether a formal roadway crossing (zebra crossing) would affect pedestrian crossing behavior and perceptions of the environment (*16*). The results showed that pedestrians felt safer and less vulnerable at a marked crosswalk. Average walking speeds were collected for pedestrians at three locations: as the pedestrian approached the crossing, across the first half of the crossing, and across the second half of the crossing. The data showed that the average walking speed increased as pedestrians entered the crosswalk where the average walking speed on the approach was 1.22 meters per second (mps) (4.0 fps) and the first half and second half average walking speeds were 1.48 mps (4.9 fps) and 1.63 mps (5.4 fps) respectively. There was an overall decrease in walking speed for all three conditions after the zebra crossing was installed; the average walking speeds for the after conditions were 1.16 mps (3.8 fps) for the approach speed and 1.31 mps (4.3 fps) and 1.39 mps (4.6 fps) for the first and second half of the crossing. While the marked crossing produced overall lower walking speeds, both conditions still showed that walking speeds increased as pedestrians entered the crosswalk. A separate study confirmed these results by studying the walking speeds of 1,792 pedestrians over the course of 18 months. The data were separated into older and younger pedestrians and it was found that both age groups walk faster when they are crossing the street compared with their normal walking speed. The walking speed for older pedestrians was 4.46 fps when they were crossing the street and 3.74 fps when they were walking normally. The younger pedestrians had a normal walking speed of 4.46 fps and 5.28 fps when crossing the street (*17*). These results compare with a crosswalk study in the state of New York that assessed behavior at primary schools and determined that when students were walking alone or in a group with a guardian their speeds were over 3.7 fps (*18*).

Manual on Uniform Traffic Control Devices

The MUTCD (Manual on Uniform Traffic Control Devices) specifies the design and installation standards for markings, striping, and signals, so a brief review as to how the MUTCD, and in particular walking speed guidance, has evolved over time is provided here for additional context. All public agencies and jurisdictions use the MUTCD to bring consistent standards to roadways, and with regard to pedestrian travel, the MUTCD has provided recommendations for walking speeds since the 1948 edition (*19*).

In the first MUTCD, quantification of pedestrian speeds was not provided as the 1948 edition only mentioned that the "safe and convenient crossing of roadway by pedestrians deserves greater attention than it has heretofore received" (*20*). In the 1961 edition, conditions in which the "normal walking speed is assumed to be 4 feet per second" and that the "WALK interval shall not be less than 7 seconds" were added. Furthermore, a provision for the pedestrian clearance interval described that the "duration should be sufficient to allow a pedestrian crossing in the crosswalk to leave the curb and travel to the center of the farthest traveled lane before opposing vehicles receive a green indication" (*21*). The 1978 version amended the WALK interval to "be at least 4 to 7 seconds in length" (*22*). In the 2000 MUTCD, the pedestrian clearance time condition was changed for travel "to at least the center of the farthest traveled lane or to a median of sufficient width for pedestrians to wait" and the defined walking speed was expanded to state that "where pedestrians who walk slower than normal, or pedestrians who use wheelchairs, routinely use the crosswalk, a walking speed of less than 1.2 meters (4 feet) per second should be considered in determining pedestrian clearance time" (*23*). In the 2003 version, a change was made regarding pedestrian clearance time as the termination location was extended to "at least the far side of the traveled way or to a median of sufficient width for pedestrians to wait" (*24*). Finally, in the current 2009 edition, the walking speed was lowered to 3.5 fps, with additional guidance stipulating that the calculation that factors in the location of the pedestrian push button utilize a walking speed of 3 fps (*25*). The calculation of the pedestrian clearance time is defined as follows:

$$PCT = D_{crosswalk} / S_{pedestrian} \quad (1)$$

where PCT is the pedestrian crossing time in seconds, $D_{crosswalk}$ is the crossing distance typically measured from

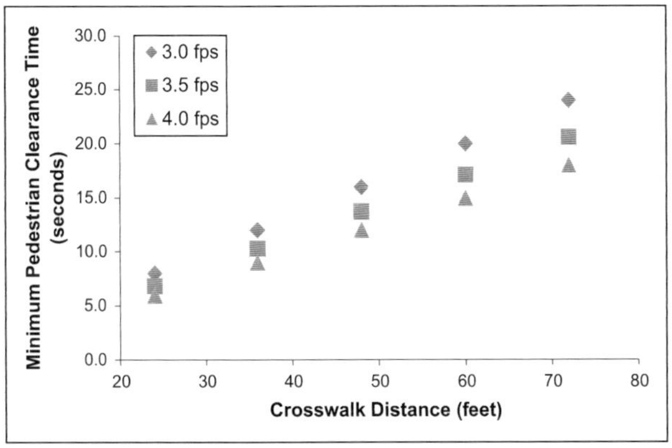

Figure 1. Pedestrian clearance time based on varying assumed walking speeds.

curb to curb along the center of the crosswalk in feet, and $S_{pedestrian}$ is the pedestrian walking speed in feet per second. This suggests that the assumed pedestrian walking speed is a defining factor that must be carefully considered.

As efforts to promote walking to school continue to evolve, local agencies may utilize enhancements in the form of pedestrian crossing flags, flashing beacons, rapid rectangular flashing beacons, or a high-intensity activated crosswalk (HAWK) beacon to create an environment where the pedestrian feels safer and parents may be less reluctant to deter their son or daughter from walking to school. In the case where a pedestrian actuated device is used, the pedestrian clearance time will be calculated based on the crosswalk distance and walking speed. As shown in Figure 1, an assumed difference in walking speed of 1.0 fps can result in an increase of up to 5 seconds (per queued vehicle) for a four-lane roadway with a two-way-left-turn lane (with a total roadway width of 60 feet). Since pedestrian crossings have an impact on vehicle traffic, if pedestrian crossing times are even slightly decreased then vehicle flow and the associated intersection delay could be improved.

Methodology

For this study, student walking speeds in marked crosswalks were collected and analyzed at seven schools in the Idaho cities of Boise, Star, and Moscow. Four elementary schools, one middle school, one high school, and one K–12 school were included as part of this study, and their locations and crosswalk crossing distances are summarized in Table 2. Figure 2 identifies the location of each school in each community (*26*, *27*).

To determine the walking speed of each school-aged child while he or she crossed a roadway, data were collected by observation and all observers were stationed at locations where the start and end point of each student crossing could be clearly identified. A distance-measuring wheel was used to measure the required distance to cross the street from the center of the crosswalk on one side of the roadway to the center of the crosswalk on the opposite side; crossing distances at a particular intersection may have varied depending on the design and width of each leg (see Table 2). The school principal was notified prior to the start of any data gathering and no student contact or engagement occurred during the study. School children were timed from the moment they stepped beyond the face of curb on one side of the street to the moment they stepped past the face of the curb on the opposite side of the street, and crossing speed was calculated by taking the time required to cross the street and dividing it by the total distance traveled. All data were collected during the 2016/17 academic year during non-inclement weather conditions and this methodology was reviewed and approved by the University of Idaho's Institutional Review Board.

It is important to note that a variety of supplemental features were present at some of these seven intersections. For example, a crossing guard guided children at Star, McDonald, Russell, and Whitmore Elementary Schools. At Moscow Middle School, recent infrastructure improvements in the form of a concrete crosswalk surface, median improvements, and crosswalk flags have created an enhanced environment for school-aged children. At Moscow High School, a flashing beacon during morning arrival and afternoon dismissal periods remind drivers that a reduced school zone speed limit is in effect when children are present. For this study it was assumed that these enhancements did not significantly affect the walking pace or behavior of students while in a given crosswalk.

Results

The tabulated results in Tables 3 and 4 were summarized to distinguish between elementary, middle, and high school students and to distinguish how the 15th, 50th, and 85th

Table 2. School Crossing Details

School	Crosswalk location	Crossing distance (feet)	Posted speed
Star	N Star Rd and W Blake Dr (north leg)	60'	30 mph
McDonald	E D St and Eisenhower St (north and east legs)	44'	25 mph
Russell	E A St and N Adams St (south and west legs)	30'	25 mph
Whitmore	E First St and S Blaine St (east leg and south legs)	24' or 30'	25 mph
Moscow Middle	E D St @ south entrance (east leg)	28'	25 mph
Moscow High	E Third St and S Adams St (east, west, and north legs)	30', 42', or 52'	25 mph
Sage	E Parkcenter Blvd and E Mallard Dr (south leg)	94'	35 mph

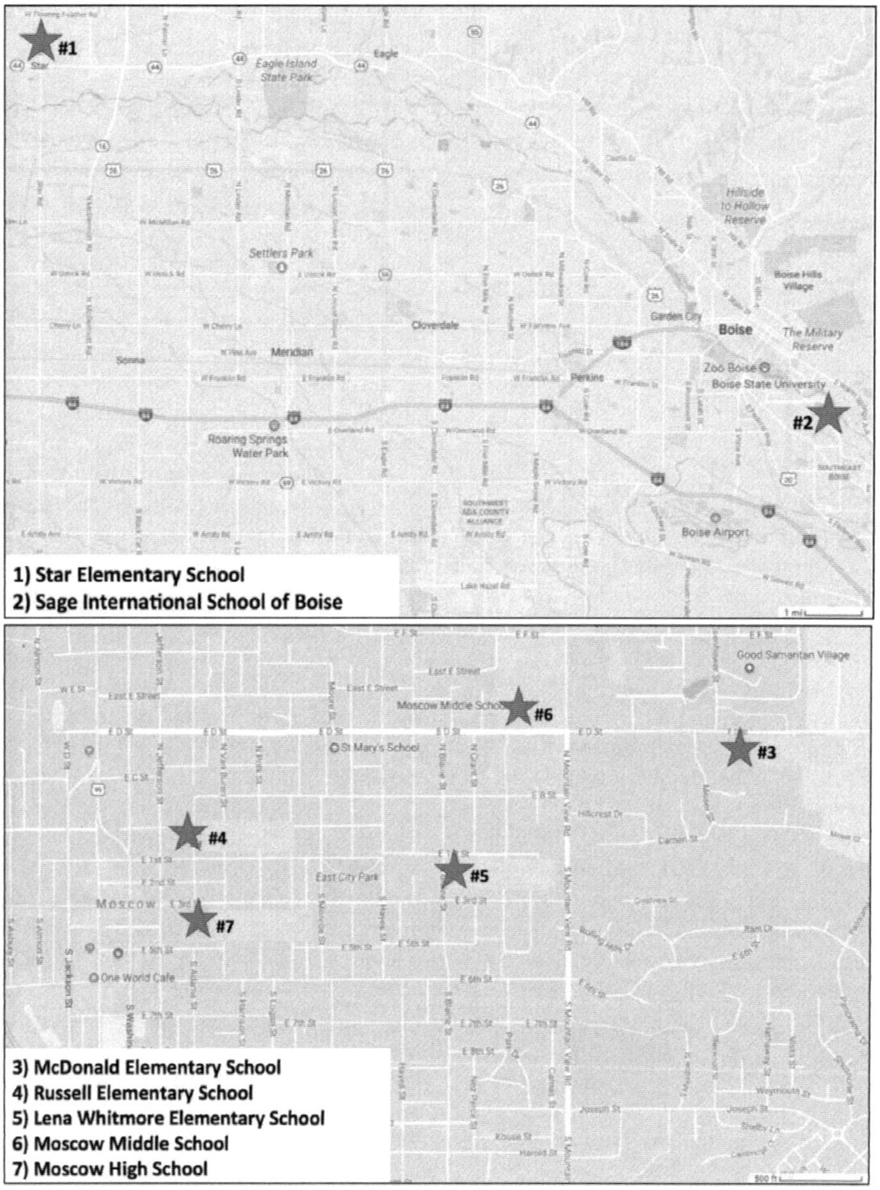

Figure 2. School crossing locations: Boise, ID and Star, ID (top) and Moscow, ID (bottom).

percentile walking speed differed at each school. All students crossing the street were assessed based on whether they were walking by themselves or in a group of two, three, four, or more than four people. Since afternoon dismissal periods represent the peak window of pedestrian travel, there were times when a student had to be subjectively identified as walking by himself or herself or as part of a group. In these cases, the student was classified as an individual unless it is abundantly clear that he or she was walking as part of a group. We categorized student walking speeds at the seven locations into three categories: all students (combined), individuals only, and students walking in a group (which represented two or more students walking together) (see Table 3). Based on these results, all of the average speeds exceeded 4.0 feet per second.

Table 4 presents the data at the individual school level and shows walking speeds at the 15th, 50th, and 85th percentiles. These percentiles are frequently used in traffic safety studies with the 15th percentile speed used to determine an appropriate walking speed for use with the implementation of signal timing. At this percentile, walking speeds at two of the elementary schools, Russell and McDonald, fell below 3.5 fps.

Table 3. Average School Walking Speeds (mph) (Individual versus Group)

Location	All			Indiv			Group		
	#	Avg	SD	#	Avg	SD	#	Avg	SD
Star	74	5.54	0.86	6	5.98	0.61	68	5.49	0.90
McDonald	46	5.98	2.43	19	5.37	2.15	27	6.95	2.61
Russell	54	4.59	1.17	17	4.73	1.20	37	4.44	1.17
Whitmore	112	4.25	0.95	33	4.42	1.17	79	4.07	0.59
Moscow Middle	83	4.84	0.89	39	5.02	0.93	44	4.47	0.67
Moscow High	213	5.01	1.17	119	5.17	1.19	94	4.55	0.99
Sage	65	6.09	1.87	26	6.56	2.13	39	5.40	1.06
All	647	5.06	1.42	259	5.19	1.45	388	4.84	1.35

Table 4. School Walking Speeds: 15th, 50th, and 85th Percentile (Individual versus Group)

	15th	50th	85th		15th	50th	85th
Star				Moscow Middle			
All	4.68	5.36	6.32	All	4.00	4.67	5.60
Indiv	5.21	6.22	6.47	Indiv	4.15	4.67	5.60
Group	4.61	5.26	6.25	Group	4.00	4.60	5.06
McDonald				Moscow High			
All	3.29	5.95	9.00	All	4.31	5.08	5.90
Indiv	3.22	4.59	8.07	Indiv	4.38	5.19	5.93
Group	4.37	6.39	9.81	Group	4.17	4.73	5.19
Russell				Sage			
All	3.38	4.41	5.95	All	4.75	5.63	7.50
Indiv	3.41	4.49	6.12	Indiv	5.30	5.73	9.09
Group	3.43	4.32	5.21	Group	4.64	4.99	6.78
Whitmore				Overall			
All	3.50	4.08	4.90	All	4.00	4.90	6.00
Indiv	3.50	4.29	6.00	Indiv	4.05	5.12	6.11
Group	3.50	4.08	4.82	Group	3.92	4.67	5.75

On average, groups of students which consisted of two, three, four, or more than four students walked slower than students who walked by themselves.

Analysis

An independent sample t-test was conducted to compare the speed of elementary, middle, and high school children walking either as an individual or in a group. A p-value, or calculated probability, was obtained from the results in which a value greater than 0.05 suggested that the mean or variance was not significantly different.

Based on the results, there was no significant difference between the cumulative walking speeds of elementary school children (mean (M) = 4.92, standard deviation (SD) = 1.55) and middle school children (M = 4.84, SD = 0.89); $t(177) = 0.42$, $p = 0.67$ or between elementary school children and high school children (M = 5.01, SD = 1.17); $t(288) = -0.62$, $p = 0.54$ (individual and group data combined). When the walking speeds of individuals were compared with those walking in groups, the age of the school child was a contributing factor. While no significant difference existed between either the walking speeds of elementary school children (M = 4.86, SD =1.53) and middle school children (M = 5.02, SD = 0.94); $t(109) = -0.72$, $p = 0.47$ or elementary school children and high school children (M = 5.17, SD = 1.19); $t(130) = -1.50$, $p = 0.14$ when only the individual data were considered, there was a significant difference between the speed of elementary school children (M = 4.97, SD = 1.59) and middle school children who walked in groups (M = 4.47, SD = 0.67); $t(70) = 2.12$, $p = 0.04$, but this significant difference was not apparent when elementary school children walking in a group (M = 4.97, SD = 1.59) were compared with high school children walking in a group (M = 4.55, SD = 0.99); $t(113) = 1.79$, $p = 0.08$. These results suggest that the interactions of school children when crossing a roadway as a group contribute to increased inattention which can result in slower crossing times and crossing time variability.

The implications from the results and analysis of this study and of previous studies are that school-aged children seem to walk with urgency while in a crosswalk and crossing a roadway. The current usage of a single value for assumed pedestrian walking speeds may unintentionally simplify how walking speeds are quantified and could encourage school-aged children to spend more time in the crosswalk environment than is necessary. The results of this study suggest that each agency should consider assessing walking speeds in their communities to determine if a more aggressive or passive assumed walking speed could be used to ensure that a fair and safe environment is created for both school-aged children who will typically cross a roadway twice a day and the motoring public who routinely travel on the same roadway.

Conclusions

This research sought to better understand school-aged pedestrian walking speeds at school crossing locations where students would be expected to be more attentive and efficient due to the potential dangers typically associated with a roadway environment and with potential pedestrian and vehicle interaction. The results suggest that current MUTCD guidelines which utilize a rate of 3.5 fps may be overly conservative, and a slightly higher assumed rate (which served as the guideline in earlier versions of the MUTCD) may in fact benefit both the driver, in the form of encountering less delay, and the school-aged pedestrian who must assume a steady yet safe walking speed when crossing the roadway in question.

Since this study focused on travel in marked crosswalks, walking speeds in unmarked crosswalks represent an area of future study. However, school-aged students are typically instructed to walk in marked crosswalks since these locations coincide with existing school walking route maps, so for these reasons only marked crosswalks were considered for this study.

The effects of inclement weather represent another area of future study. It is hypothesized that during rainy, windy, or snowy conditions school-aged children would walk with more urgency en route to and from school, but these effects need to be further examined.

This study did not examine the walking speed of school-aged children who were confined to a wheelchair or required the support of crutches or other walking aids. Younger children, particular those of elementary school age, likely would not yet have the arm strength to use these devices.

It was observed that some users did not stay within the parameters of the crosswalk, which could have resulted in a small reporting error. Some students started on the perimeter of the delineated crosswalk area and slowly worked their way along the centerline of the crosswalk after a few paces. Because the start and stop location of the occupants was identified as the face of curb on opposite sides of the roadway, children who veered from this defined path would have traveled a slightly longer distance than that measured for the crosswalk distance, resulting in a lower walking rate since one crosswalk distance was retained for all calculated values.

Future studies could expand the framework of this study to explore how specific socioeconomic factors such as gender, race, and family income may also be contributors to actual walking speeds. The additional data collected from each school child could then be used to apply more robust analytical methods that accounted for the increased number of explanatory variables.

It should also be noted that this study has focused on school-aged children since the operation of any devices would likely be used in conjunction with morning arrival and afternoon dismissal periods in the vicinity of a school. At non-exclusive school locations, all ages and population groups should be considered, from young children to elderly users, along with those with disabilities. These devices serve all pedestrians, including school children, but the most vulnerable population groups often benefit the most, and cross at a slower pace.

With approximately 6% of trips in the United States attributed to pedestrian travel (*28*), it is evident that the percentage of walk-to-school trips by young children, despite its decades-long decline, continues to outpace the total percentage of overall trips made by adults. A representative from Star Elementary School noted that approximately 15% of students currently walk to school and use the crosswalks near the school, matching the current national percentage. If an increase of walking is to be expected at Star Elementary and at other elementary, middle, and high schools, a safe walking environment must be provided to create a foundational starting point, while the efficient and effective use of certain devices will serve to further enhance this environment. The pedestrian walking speed that is used to establish the operational duration of a safety device must be appropriately assumed. In turn, pedestrians would be expected to comply with and react and respond to these devices in a reasonable and prudent manner. A fair balance will allow for the effective management of both the school-aged pedestrian and the motoring public to ensure that school children, with the permission of their parents, are more willing to walk to school while motorists will acknowledge crossing pedestrians when they are present and be conditioned to acknowledge these devices for their full operational duration.

Author Contributions

The authors confirm contribution to the paper as follows: study conception and design: KC; data collection: PF, ML, EB; analysis and interpretation of results: KC, PF, ML, EB; draft manuscript preparation: KC, PF, ML, EB. All authors reviewed the results and approved the final version of the manuscript.

References

1. National Center for Safe Routes to Schools. *How Children Get to School: School Travel Patterns From 1969 to 2009.* National Center for Safe Routes to Schools, Chapel Hill, NC, 2011, pp. 2.
2. *Bicycling and Walking in the United States: Benchmarking Report.* Alliance for Biking and Walking, Washington, D.C., 2016, pp. 12.
3. The State of Obesity in Idaho. *The State of Obesity: Better Policies for a Healthier America.* Trust for America's Health and the Robert Wood Johnson Foundation, Washington, D.C., stateofobesity.org/states/id/. Accessed June 25, 2017.
4. Knutson, L. M., P. A. Schimmel, and A. Ruff. Standard Task Measurements for Mobility: Thirty-Second Walk Test. *Pediatric Physical Therapy*, Vol. 11, 1999, pp. 183–190.
5. David, K. S., and M. Sullivan. Expectations for Walking Speeds: Standards for Students in Elementary Schools. *Pediatric Physical Therapy*, Vol. 17, 2005, pp. 120–127.

6. Sutherland, D., R. Olshen, E. Biden, and M. Wyatt. *The Development of Mature Walking*. Cambridge University Press, London, 1988.
7. Cristina de David, A., C. Esteves, T. Nunes, P. Dini, and E. Manfio, E. *Walking and Running Spatio-Temporal Variables in Children from Three to Six Year Old*. University of Ottawa, Ottawa, Canada. health.uottawa.ca/biomech/csb/Archives/csb2006.pdf. Accessed June 26, 2017.
8. Stansfield, B. W., S. J. Hillman, M. E. Hazlewood, and J. E. Robb. Regression Analysis of Gait Parameters with Speed in Normal Children Walking at Self-Selected Speeds. *Gait and Posture*, Vol. 23, No. 3, 2006, pp. 288–294.
9. Dusing, S. C., and D. E. Thorpe. A Normative Sample of Temporal and Spatial Gait Parameters in Children Using the GAITRite® Electronic Walkway. *Gait and Posture*, Vol. 25, No. 1, 2007, pp. 135–139.
10. Lythgo, N., C. Wilson, and M. Galea. Basic Gait and Symmetry Measures for Primary School-Aged Children and Young Adults Whilst Walking Barefoot and with Shoes. *Gait and Posture*, Vol. 30, No. 4, 2009, pp. 502–506.
11. Knoblauch, R., M. Pietrucha, and M. Nitzburg. Field Studies of Pedestrian Walking Speed and Start-Up Time. *Transportation Research Record: Journal of the Transportation Research Board*, 1996. 1538: 27–38.
12.Gates, T. J., D. A. Noyce, A. R. Bill, and N. Van Ee. Recommended Walking Speeds for Pedestrian Clearance Timing Based on Pedestrian Characteristics. *Transportation Research Record: Journal of the Transportation Research Board*, 2006. 1982: 38–47.
13.Mitman, M., and D. Ragland. Crosswalk Confusion: More Evidence Why Pedestrian and Driver Knowledge of the Vehicle Code Should Not Be Assumed. *Transportation Research Record: Journal of the Transportation Research Board*, 2007. 2002: 55–63.
14Knoblauch, R., M. Nitzburg, and R. Seifert. Pedestrian Crosswalk Case Studies: Richmond, Virginia; Buffalo, New York; Stillwater, Minnesota. Report No. FHWA-RD-00-103. Federal Highway Administration, Washington, D.C., 2001.
15.Knoblauch, R., and P. Raymond. *The Effect of Crosswalk Markings on Vehicle Speeds in Maryland, Virginia, and Arizona*. Report No. FHWA-RD-00-101. Federal Highway Administration, Washington, D.C., 2000.
16. Harvard, C., and A. Willis, A. Effects of Installing a Marked Crosswalk on Road Crossing Behaviour and Perceptions of the Environment. *Transportation Research Part F: Traffic Psychology and Behavior*, Vol. 15, 2012, pp. 249–260.
17Montufar, J., J. Arango, M. Porter, and S. Nakagawa. Pedestrians' Normal Walking Speed and Speed When Crossing a Street. *Transportation Research Record: Journal of the Transportation Research Board*, 2007. 2002: 90–97.
18. Yang, W., H. Li, H. J. Park, W. Yu, S. Li, and S. Park. Crossing Behavior and Optimal Signal Control Policy at Primary School Crosswalks. Presented at 95th Annual Meeting of the Transportation Research Board, Washington, D.C., 2016.
19LaPlante, J., and T. Kaeser. The Continuing Evolution of Pedestrian Walking Speed Assumptions. *Institute of Transportation Engineers, ITE Journal*, Vol. 74, No. 9, 2004, pp. 32–40.
20. *Manual on Uniform Traffic Control Devices*. Public Roads Administration, Washington, D.C., 1948.
21. *Manual on Uniform Traffic Control Devices*. Bureau of Public Roads, Washington, D.C., 1961.
22. *Manual on Uniform Traffic Control Devices*. United States Department of Transportation: Federal Highway Administration, Washington, D.C., 1978.
23. *Manual on Uniform Traffic Control Devices*. United States Department of Transportation: Federal Highway Administration, Washington, D.C., 2000.
24. *Manual on Uniform Traffic Control Devices*. United States Department of Transportation: Federal Highway Administration, Washington, D.C., 2003.
25. *Manual on Uniform Traffic Control Devices*. United States Department of Transportation: Federal Highway Administration, Washington, D.C., 2009.
26. Google Maps. *Boise, Idaho*, 2017. https://www.google.com/maps/@43.5884125,-116.3228265,12z.
27. Google Maps. *Moscow, Idaho*, 2017. https://www.google.com/maps/@46.7350388,-116.9913989,15z.
28. Pucher, J. Promoting Safe Walking and Cycling to Improve Public Health: Lessons from the Netherlands and Germany. *American Journal of Public Health*, Vol. 93, No. 9, 2003, pp. 7.

The Standing Committee on Transportation Safety Management (ANB10) peer-reviewed this paper (18-00688).

Article

TRR
JOURNAL OF THE TRANSPORTATION RESEARCH BOARD

Transportation Research Record
2018, Vol. 2672(32) 30–39

DOI: 10.1177/0361198118794285
journals.sagepub.com/home/trr
SAGE

Development of a Crash Risk-Scoring Tool for Pedestrian and Bicycle Projects in Oregon

Yi Wang[1], Christopher M. Monsere[1], Chen Chen[2], and Haizhong Wang[2]

Abstract

Methods for identifying and prioritizing high-crash locations for safety improvements are generally crash-based. There are fewer reported crashes involving non-motorized users and, in most states, reported crashes must involve a motor vehicle. This means that minor, non-injury events are not reported and those crashes that are reported tend to be more severe. Selecting projects based only on crash performance is sometimes limiting for these crash types and predicting where these crashes will occur next is also a challenging task. An alternative to crash-based selection is to develop risk-based criteria and methods. This paper presents the results of a research effort to develop a risk-scoring method with weights derived from data for use in project screening and selection in Oregon. To develop the risk model, data were collected from 188 segments and 184 intersections randomly selected on both state and non-state roadways. Geometric, land use, volume, and crash data were collected from Google Earth, EPA's Smart Location Database, and the Oregon Department of Transportation crash database from 2009 to 2013. The sample included 213 bicycle and pedestrian crashes on the segments and 238 at intersections. Logistic regression models were developed and the outputs used to create pedestrian and bicycle risk-scoring tools for segments and intersections. The risk-scoring tool was applied to safety projects identified in the 2015 All Roads Transportation Safety (ARTS) project lists from Oregon. The risk scores for the case study applications aligned reasonably well with the project's benefits-costs estimates.

Methods for identifying and prioritizing high-crash locations for safety improvements are generally based on reported crashes (*1*). There are fewer reported crashes involving non-motorized users and, in most states, reported crashes must involve a motor vehicle. This means that minor, non-injury events are not reported and those crashes that are reported tend to be more severe. This results in fewer crashes for network screening techniques to identify locations. Further, since there is clear evidence that the decision to make non-motorized trips by bicycling or walking is influenced by the perception of safety (*2*), locations that have deficient geometric or operational features may not be identified by crash-based screening methods. Exposure data for non-motorized users is also challenging to obtain at the network level (*3*). Selecting projects based only on crash performance is sometimes limiting for these crash types and predicting where these crashes will occur next is also a challenging task. An alternative to crash-based selection is to develop risk-based criteria and methods.

The Oregon Department of Transportation (ODOT) has identified pedestrian and bicycle crashes as a primary focus area for investing in infrastructure funding. In 2015, 17.7% of the 445 traffic fatalities in Oregon were non-motorized (*4*). ODOT has appropriated approximately $4 million annually in the ARTS program to address this issue. Prior to this research project, ODOT conducted a systemic safety analysis of pedestrian and bicycle safety (*5*). As part of the work, a crash frequency-based and risk-based prioritization methodology were developed. The quantification of risk factors and the magnitude of their influence was constrained by the additional data that the project was able to collect, and many of the risk scores were based on engineering judgment.

The objective of the research described in this paper was to develop a risk-scoring method with weights

[1]Department of Civil and Environmental Engineering, Portland State University, Portland, OR
[2]School of Civil and Construction Engineering, Oregon State University, Corvallis, OR

Corresponding Author:
Address correspondence to Christopher M. Monsere: monsere@pdx.edu

derived from data. The intent is for the risk method to be used in project screening and selection in Oregon. Following a brief background and literature summary, the data collection methodology on segments and at intersections is described. In the methodology, the logistic modeling approach and conversion of the outputs to a risk-scoring tool is described. The model results and final tool are then described. Finally, the application of the risk-scoring tool to several projects recommended for funding in the ARTS project list is presented.

Background

In the transportation context, risk is defined as a probability or threat of damage, injury, liability, loss, or any other negative occurrence that is caused by external or internal vulnerabilities, and that may be avoided through preemptive action. The amount of risk can be interpreted by the probability of the outcome and potential severity of the outcome if the event occurs (*6*). For transportation and vulnerable road users, the probability is a function of exposure and consequence is a function of operating conditions (e.g., vehicle speeds and size). Risk assessment and scoring methods should include elements of exposure and expectations of the severity of the outcome.

In the general sense, the systemic safety approach has some of the elements of a risk-based assessment. After first identifying a focus crash type, a more detailed analysis can identify, diagnose and treat locations that are at high risk for crashes on a system-wide basis (*7*). More directly related are the road assessment programs such as the United States Road Assessment Program (usRAP) (*8*). The program categorizes roadways in a number of traditional ways (frequency and rate of fatal crashes, the difference from average rate performance), but also includes a method to develop a road protection score. The road protection score is calculated based on the potential for severe outcomes for head-on, run-off-the-road, and intersection crashes. In this way, the assessment is not entirely crash-based and includes factors that explain the probability of a severe crash.

For non-motorized project selection and prioritization, the Pedestrian and Bicycle Safety Indices (ISI) developed for the Federal Highway Administration (FHWA) are similar in intent to a risk-scoring model. The indices allow engineers and planners to identify proactively which intersection crossings and approach legs should be the greatest priority for undergoing pedestrian and bicycle safety improvements (*9*). In the methodology, the ISI score is an evaluation of each approach leg of an intersection rather than evaluating the intersection as a whole. Safety ratings (opinion) from experts and bicycle/pedestrian-motorist interactions from a video analysis of each site were used to generate a multivariate linear regression model to explain the safety indices. More recently, the ActiveTrans Priority Tool (*10*), produced as part of National Cooperative Highway Research Program Report 803, allows users to select scoring criteria and input the weights of stakeholder input, constraints, and opportunities, and to prioritize the locations. While not derived from crash analysis, level of service measures such as the *Highway Capacity Manual's* multimodal level of service for pedestrian and bicycle facilities generally measure user comfort and convenience. These geometric and operational features have some relationship to safety as demonstrated in the literature. For bicycles, the level of traffic stress (*11*) is a tool only for bicycle networks. This method estimates the stress level by the criteria based on Dutch standards for bicycle facilities. Preliminary studies show that more than half of the bicycle crashes happened on streets with a higher level of stress (*12*).

Finally, there is a large body of literature in which the severity and frequency of pedestrian and bicycle crashes have been explored. With respect to severity and crash probabilities, many studies (*13–20*) have used a logit model to identify the significant variables, including geometric design characteristics, driver characteristics, and built environment variables, in pedestrian and bicycle crashes at both road segment and intersections.

Data

The previous literature was used to identify key variables that should be considered in a risk model. Tables 1 and 2 summarize the list of variables that were collected and their descriptive statistics. A random sampling approach was used to select segments and intersections for data collection. Segments were selected if they were in an urban area and had arterial (minor or principal) functional classification. Both state- and non-state-owned roadways were included but ramps were excluded. A selected segment was split to be homogenous in the data elements (but only one segment was kept to minimize spatial correlations). Intersections were collected concurrently with the segment sampling process if the selected segment contained traffic control on the major road (stop or signal).

A total of 188 arterial roadway segments and 184 intersections with traffic control (traffic signal, four-way stop, or roundabout) within urban areas were randomly selected. The average length of segments is roughly 706 feet. The elements of geometric design were primarily collected manually from inspection and measurement of Google Earth aerial photos, and the built environment characteristics were gathered from the U.S. Protection Agency's Smart Location Database. Traffic volume data were assembled for all segments and intersections, and

Table 1. Summary of Categorical Variables

Variable	Frequency	Percentage of sample
Segments		
Number of traffic lanes (excluding two-way left-turn lane)		
1 Lane	2	1.06
2 Lane	112	59.57
3 Lane	16	8.51
4 Lanes or more	58	30.85
Presence of two-way left-turn lane (TWLTL)		
No	130	69.15
Yes	58	30.85
Presence of marked midblock crosswalks within segment		
No	181	96.28
Yes	7	3.72
Presence of on-street parking		
No	143	76.06
Yes	45	23.94
Traffic direction		
One-way	24	12.77
Two-way	164	87.23
Posted speed limit (mph)		
20	12	6.38
25	40	21.28
30	27	14.36
35	79	42.02
>35	30	15.96
Intersections		
Intersection legs		
4-Leg	157	85.33
3-Leg	27	14.67
Major road. Presence of right-turn lane		
No	133	72.28
Yes	51	27.72
Major road. Presence of median		
No	165	89.67
Yes	19	10.33
Minor road. Functional class		
Arterial	92	50.00
Collector	92	50.00
Minor road. Presence of right-turn lane		
No	120	65.22
Yes	64	34.78
Minor road. Total number of traffic lanes		
2	43	23.37
3	90	48.91
4	26	14.13
>4	25	13.59
Number of transit lines that go through the intersection		
0	34	18.48
1	90	48.91
2	49	26.63
3	9	4.89
>3	2	1.09

Table 2. Summary of Continuous Variables

Variable	Mean	Standard deviation
Total population density (people per square mile)	4086.92	5333.08
Three-leg intersection density (per square mile)	162.66	110.12
Number of transit lines through intersection	1.36	1.18
Major road AADT	14,080.34	8,143.53
Minor road AADT	7,648.98	5,480.24
Segment AADT	10,806.27	7,607.40
Bicycle volume, intersection	261.26	641.24
Bicycle volume, segment	94.32	165.37

all of the annual average daily traffic (AADT) data were calibrated to the year 2014 using growth factors obtained from the ODOT ATR station growth factors. STRAVA Metro data was used to represent average daily bicycle traffic in this study. The dataset was purchased by ODOT for research and project purposes (OSU and PSU researchers were allowed to obtain the data under this agreement). Unpublished work by ODOT suggests that the STRAVA count can represent 1% of total bicycle volume without considering the difference between commuter and cyclist.

Crash data for five years, from 2009 to 2013, in Oregon was used in this study. Crashes within a 125-foot radius buffer zone at the middle point of intersections were considered as intersection-related crashes. In total, there were 113 pedestrian crashes and 100 bicycle crashes on the segments, and 108 pedestrian crashes and 130 bicycle crashes at the intersections. As designed, the sample produced segments and intersections where there have been no crashes. As shown in Figure 1, around 20% of the segments have only one bicycle or pedestrian crash. Fewer segments, around 10%, have two crashes in five years and only a few have more than two crashes. Intersections have similar patterns for pedestrian and crash frequency. Most of the intersections selected do not have crashes, and few intersections had more than two crashes from 2009 to 2013. Others have one or two pedestrian or bicycle crashes.

Methodology

Logistic regression was initially proposed by Cox in 1958 (*21*) to measure the categorical dependent variable (Y) and multiple independent variables (X) by using the logistic function. It was developed based on the idea of odds which describes likelihoods of events. Specifically, the odds indicate how often something (e.g., y = 1) happens relative to how often it does not happen (e.g., y = 0) (*22*). When developing the logistic regression equation,

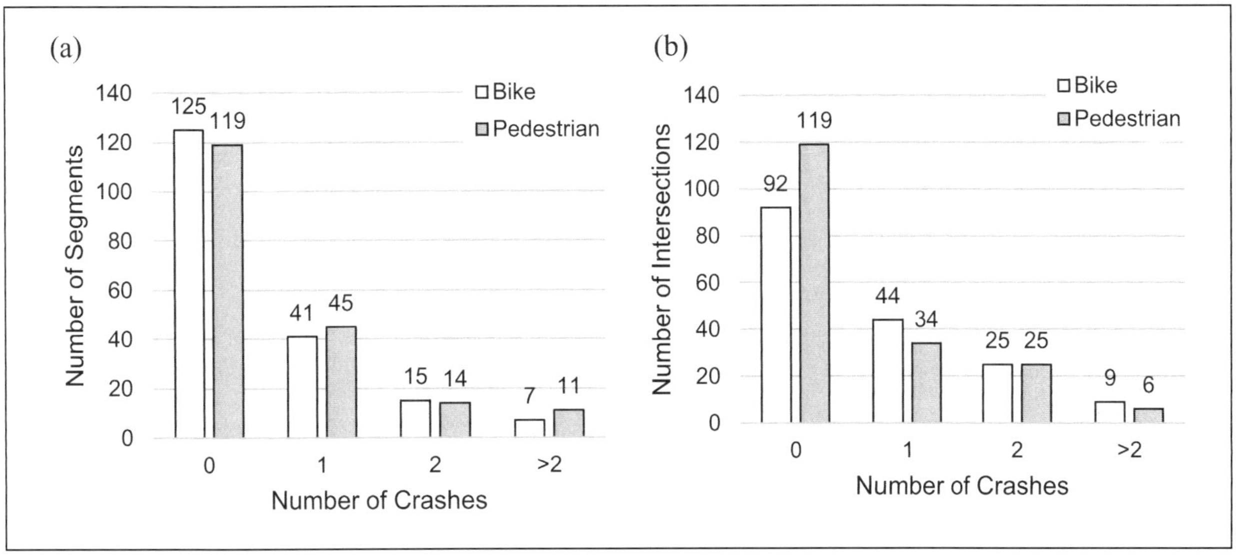

Figure 1. Number of segments (*a*) and intersections (*b*) by crash frequency.

the *ln* of odds represents a logit transformation, where the logit is a function of covariates (*23*):

$$logit[P_n] = \ln\left[\frac{P_n}{1-P_n}\right] = \beta_0 + \beta_1 x_{1,n} + \cdots + \beta_i x_{i,n} \quad (1)$$

Those parameters, after estimation by maximum likelihood methods, can be used to estimate the probability that the outcome takes the value 1 as a function of covariates using the equation below (*21*, *24*):

$$P_n = \frac{e^{(\hat{\beta})}}{1 + e^{(\hat{\beta})}} \text{ where } \hat{\beta} = \beta_0 + \beta_1 x_{1,n} + \cdots + \beta_i x_{i,n} \quad (2)$$

where P_n is the probability that an event (crash) happens for observation n (indicates a segment or an intersection in this paper). β_0 is the model constant and $\beta_1 \cdots \beta_i$ are the unknown parameters corresponding with the independent variables (total i). e is the exponential constant approximately equal to 2.71828; x is a vector of independent variables and $\hat{\beta}$ is a vector of estimated parameters; i is the total number of independent variables. This study uses the binary variable that whether bicycle crashes occurred or not on a site as the response variable and is called "crash occurrence model" thereafter. Crash frequency is transferred into crash occurrence (crash occurrence = 1 when there was more than one crash happened; and = 0 the otherwise) and served as a binary response variable.

The model coefficient cannot be used directly to interpret the slope or rate of change of the dependent variable per unit change in the independent variable in logit model, however, due to its nonlinear feature. The logit model is developed based on odds that describe likelihoods of events. Odds, in Equation 3, are related to probability but are conceptually and numerically different (*21*).

$$Odds = \frac{P_n}{1-P_n} = e^{\beta_0 + \beta_1 x_{1,n} + \cdots + \beta_i x_{i,n}} \quad (3)$$

Meaningful interpretation of coefficients in logistic models relies on the interpretation of the difference between two odds (*25*). Equation 2 shows that the logit model is linear in the logit. Especially for a unit change in x_i, we expect the logit change by β_i, holding all other variables constant. The odds ratio, which is shown in Equation 4, compares the likelihood of two odds and provides a meaningful interpretation of coefficients. If the odds ratio is greater than 1, it represents the likelihood of an event with changing one unit in one variable (indicated in the numerator) is greater than the likelihood of the event with original value in variables (indicated in the denominator).

$$Odds\ Ratio = \psi = \frac{odds\ of\ an\ event\ with\ increase\ one\ unit\ in\ one\ variable}{odds\ of\ an\ event} = \frac{\left(\frac{P_n(i)}{1-P_n(i)}\right)^*}{\left(\frac{P_n(i)}{1-P_n(i)}\right)} = \frac{e^{\beta_0 + \beta_1 x_{1,n} + \cdots + \beta_i (x_{i,n}+1)}}{e^{\beta_0 + \beta_1 x_{1,n} + \cdots + \beta_i x_{i,n}}} = e^{\beta_i} \quad (4)$$

Table 3. Pedestrian Modeling Results

Variable	Coefficients	Standard error	P-value	Significance	Odds ratio
Segments					
Travel direction (one-way or two-way)	−1.289	1.018	0.001	***	0.276
Presence of on-street parking	1.337	0.514	0.012	*	3.808
Presence of two-way left-turn lane	1.071	0.384	0.005	**	2.918
Posted speed limit (mph)	0.047	0.027	0.082	.	1.048
Total population density (people per square mile)	0.0017	0.00006	0.006	**	1.002
Number of traffic lanes (excluding two-way left-turn lane)	0.370	0.016	0.023	*	1.447
Null deviance: 249.16 on 188 degrees of freedom. Residual deviance: 204.62 on 182 degrees of freedom AIC: 218.62					
Intersections					
Total population density (people per square mile)	0.00024	0.000072	0.000	***	1.000
Number of transit lines through intersection	0.383	0.208	0.065	.	1.467
Major road, number of right-turn lanes	0.784	0.432	0.070	.	2.190
Major road, AADT 2014	0.000063	0.000023	0.005	**	1.000
Major road, presence of median	−1.260	0.664	0.058	.	0.284
Minor road, number of right-turn lanes	−1.312	0.440	0.003	**	0.269
Null deviance: 238.99 on 183 degrees of freedom Residual deviance: 195.40 on 177 degrees of freedom AIC: 209.4					

Note: Significance code: 0 ***; 0.001 **; 0.01 *; 0.05 ".".

where $\left(\frac{P_n(i)}{1-P_n(i)}\right)^*$ is the odds that an event with changing one unit in one variable (i.e., $x_{i,n}+1$) happens whereas $\left(\frac{P_n(i)}{1-P_n(i)}\right)$ is the odds that the event with original value in variables happens. e^{β_i} is the odds ratio that indicates the relative amount by which the odds of an outcome change when the value of a corresponding independent variable increases by one unit (*22*).

The model results were then converted to a risk score based on the odds ratio. Conversion to a risk score creates an easy method for interpreting and applying the modeling results. For each type of facility and user, the maximum risk score is set to 100. The distribution of 100 points to each of the variables was weighted. Additional details on this process can be found in the project's final report (*25*). The risk scores are comparable for a facility and user type but not across the categories (i.e., the scores from the pedestrian segments are not comparable to the bicycle intersections). Comparison across the categories can be done by estimating the risk score percentile (discussed in the application section).

Results

To develop the logit models, a combined backward and forward stepwise method (*26*) was initially used to determine the significant variables to be included in the model, then other possible variables (based on engineering judgment and literature) were also explored. The dependent variable was the presence of a crash or not (0 or 1). Table 3 summarizes the final selected model for pedestrian segments and intersections. Table 5 translates the model parameters into the risk score. In this modeling effort, the significant variables should not be interpreted as recommendations for engineering-level improvements. The variables, in many cases, explain more about the safety of the location than the individual variable. Design-level safety decisions should use more robust tools such as the *Highway Safety Manual* (*1*).

For pedestrian segments, the significant variables are reasonable and the direction of effect is as expected. The model includes variables of exposure and risk. The odds ratio for the travel direction (one-way to two-way) indicates that the odds of a pedestrian crash happening on a segment are 0.276 times smaller, holding all other variables constant. These results likely reflect some association with more traffic and lanes on one-way streets. The presence of parking increases the odds of pedestrian crashes (more pedestrian activity). The effect of increased posted speeds on the probability of crashes is in the expected direction (an increase in speed increases the probability of a crash outcome). The continuous nature of how the speed was modeled also means that the effect is not linear, and that higher speeds have a very significant effect on the overall probability prediction. The presence of a two-way left-turn lane (TWLTL) would also increase crash potential. The hypothesis is that the additional width and turning conflicts add additional risk for pedestrians or higher volumes associated with TWLTL roads. As a measure of pedestrian activity, the total population density coefficients indicate that crash probability increases with population density.

For pedestrians at intersections, the final model includes total population density; the number of transit lines through the intersection; the number of major-road right-turn lanes; the major-road AADT in 2014; the presence of a median on the minor road; and the number of

Table 4. Bicycle Modeling Results

Variable	Coefficients	Standard error	P-value	Significance	Odds ratio
Segments					
Presence of marked midblock crosswalks within segment	−1.207	−3.662	0.000976	***	0.2991
AADT 2014	0.00003187	0.00002124	0.1336		1.00003187
Three-leg intersection density (per square mile)	0.002087	0.001486	0.1602		1.002089
Bicycles per day (STRAVA)	0.001007	0.000994	0.3110		1.0010012
Null deviance: 240.60 on 188 degrees of freedom Residual deviance: 220.35 on 183 degrees of freedom AIC: 230.35					
Intersections					
Bicycles per day (STRAVA)	0.00146	0.00024	0.0368	*	1.001
Number of transit stops	0.3507	0.1924	0.0683	.	1.420
Minor functional class (arterial as base)	−0.9096	0.3585	0.0112	*	0.4027
Minor road, total number of traffic lanes	0.49698	0.2067	0.0231	*	1.644
Minor road, presence of right-turn lane	−0.7056	0.3581	0.0488	*	0.4938
Null deviance: 232.04 on 167 degrees of freedom Residual deviance: 200.04 on 161 degrees of freedom AIC: 214.04					

Note: Significance code: 0 ***; 0.001 **; 0.01 *; 0.05 ".".

right-turn lanes on the minor road. The density and transit variables reflect pedestrian activity levels and the direction is as expected (increased probability with increases in these measures). Similarly, the coefficient of major-road AADT was significant and positive. The presence of right-turn lanes on the major road is associated with an increased crash probability. The lack of a median on the major road also increases the crash probability. However, on the minor roadway, the coefficient for the presence of a right-turn lane on the minor road is not as expected.

Table 4 summarizes the final selected model for bicycle segments and intersections, including model coefficients, standard error, p-value, significance, and odds ratios. Table 6 translates the model parameters into the risk score. Unlike the pedestrian models, few variables were found to be significant in the bicycle models, especially for the segment model. For the bicycle intersection models, the significant variables include bicycles per day, the number of transit stops, the minor-road functional class, minor-road total traffic lanes, and minor-road right-turn lanes. Clearly, the number of bicycles per day captures the increased exposure as volumes increase. The number of transit stops indicates a presence of other road users and possibly additional interactions with bus traffic. The number of lanes on the minor road can be interpreted as increasing the total intersection size.

For the bicycle segment model, the significant variables include the presence of a crossing (no crossing is the base condition); AADT (factored to 2014); three-leg intersection density; and bicycle volume (per day). Unlike all the other crash occurrence models, very few variables were found to be significant in the model. The final selected model includes variables at a lower confidence level for significance than the other models. The final model does include an exposure metric for bicycles per day and the sign is expected (as bicycles per day increase then probability also increases). Three-leg intersection density (hypothesized to be associated with less connectivity) is associated with a positive increase in crash probability. Vehicle volume, represented by AADT in the model, has the positive coefficients as expected, indicating that high vehicle volume could lead to high risk for bicyclists on this segment. The presence of crossings decreases the probability of bicycle crashes on the segments. It is hypothesized that the presence of pedestrian crossings is related to the overall design of the roadway (i.e., a more non-motorized, user-friendly character).

Application

The risk-scoring tables were applied to all segment and intersection samples in the modeling dataset. Figure 2 shows the distribution of scores for each of the risk tools as applied to these locations, and the percentile and mean values of these distributions. With the exception of the intersection tool for pedestrians, the distributions are skewed. The bicycle segment model is less distributed, reflecting the limited number of significant variables that were in the model. The risk scores are only intended to be evaluated within each context (i.e., the risk score for bicycles at intersections is not comparable to the score for pedestrians on segments). The distributions of the risk scores could be used when making comparisons across the tools by estimating the percentile of the score. The values for the 25th, 50th and 75th percentile are shown in Figure 2. For example, a risk score of 46 would be above the 75th percentile of the calculated scores for the pedestrian segment and bicycle intersection risk scores, but only average for the other two tools.

To further demonstrate the application of the tool, a total of 10 intersection projects was selected from the ARTS final project list in the more urban ODOT regions.

Table 5. Summary of Risk Scores for Pedestrians

Segments			Intersection		
Variables	Levels	Risk score	Variables	Levels	Risk score
Total population density (per square mile)	<=1000	0	Total population density (per square mile)	<=1000	0
	1001–3000	6		1001–3000	5
	3001–5000	8		3001–5000	8
	5001–7000	11		5001–7000	13
	>7000	20		>7000	21
Traffic direction	One-way	17	Number of transit lines with routes through intersection	0 (base)	0
	Two-way	0		1	6
On-street parking	Yes	17		2	8
	No	0		3	12
Posted speed limit (mph)	<=25	0		>3	25
	30	6	Major AADT (2014)	<=5000	0
	35	8		5001– 10000	5
	>35	12		10001– 15000	7
Presence of TWLTL	Yes	14		15001– 20000	10
	No	0		20001– 25000	13
Total traffic lanes	2	0		>25000	18
	3 or 4	10	Presence of median on major road	Yes	0
	>4	20		No	13
			Minor road, presence of right-turn lanes	Yes	0
				No	15
			Major road, presence of right-turn lanes	No	0
				Yes	8

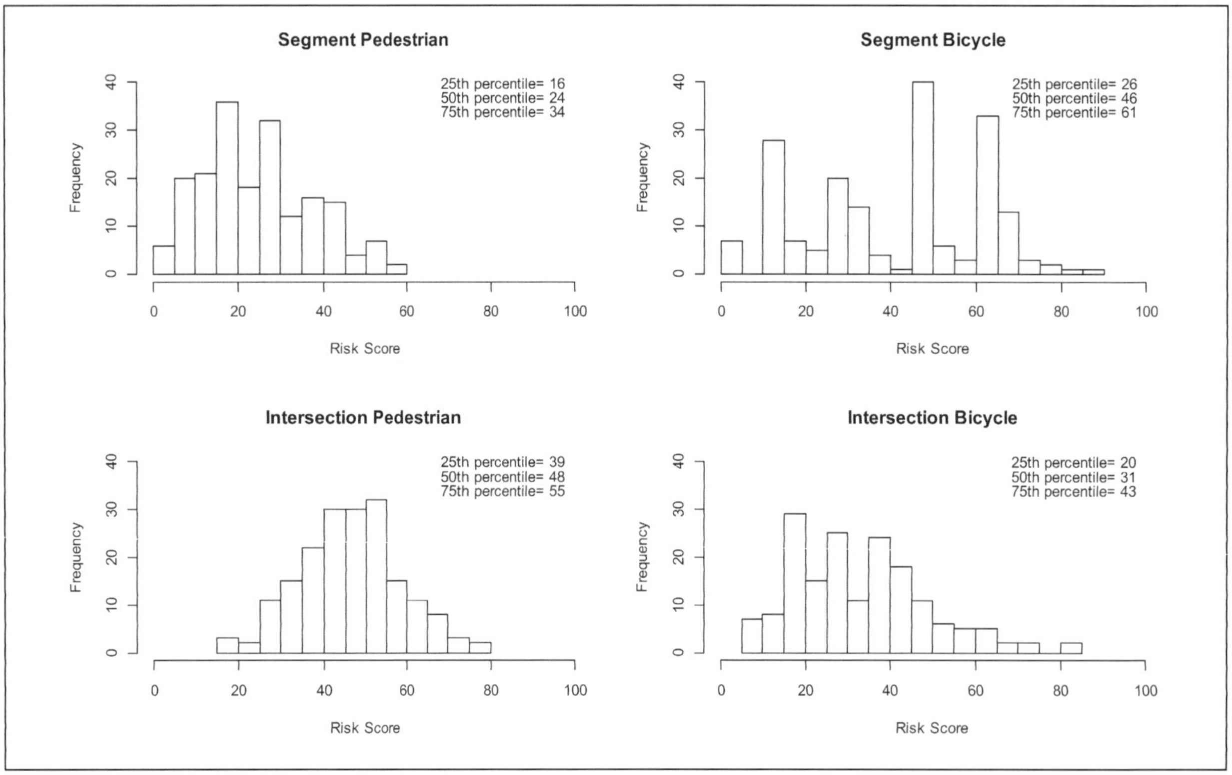

Figure 2. Risk score distribution for each facility type and user.

Table 6. Summary of Risk Scores for Bicycles

Segments			Intersections		
Variables	Levels	Risk score	Variables	Levels	Risk score
Bikes per day (STRAVA)	<=200	0	Bikes per day (STRAVA)	<=200 (base)	0
	201–800	15		<=800	11
	>800	25		>800	20
AADT	<=5000	0	Number of transit stops	0 (base)	0
	5001–10000	12		1	7
	10001–15000	14		2	10
	15001–20000	16		3	14
	20001–25000	19		>3	27
	>25000	25	Minor functional class	Collector	0
Three-leg intersection density	1–150	0		Arterial	12
per square mile	151–200	13	Minor road total number of traffic lanes	2 (base)	0
(EPA Smart Location)	>200	16		3	8
Presence of marked crosswalk	Yes	0		4	12
	No	34		>4	31
			Minor road presence of right-turn lane	Yes (base)	0
				No	10

The projects are listed in Table 7, including brief descriptions of the proposed project elements. Table 7 also includes the benefit-to-cost ratio (B/C) of each of the projects as reported by ODOT. To calculate the risk score, the values for each of the variables at the project location were collected, then the risk scores for each variable (shown in Table 3 (for pedestrians) and Table 5 (for bicycles)) were summed to compute a project risk score. As a reminder, the risk scores between the two types are not comparable. Therefore, the risk score percentile is also shown in Table 7.

As the projects listed in Table 7 were the final projects selected through the ARTS process, it is expected that the projects with higher B/C value also have higher risk scores. The application of the risk scores corresponds well to the project's final B/C value. Referring to Table 7, the risk scores for the projects with the two highest B/C ratios are above the 75th percentile score of pedestrian intersection models with scores of 55. The lower-ranked B/C projects correspond to lower percentiles of the risk score. A similar exercise was conducted for another five intersection projects that relate to bicycles. As shown, the higher risk scores align with the higher B/C ratios with the exception of the Albany project. The 75th percentile risk score is 42.75; all of the projects are below this value.

Conclusion

In this study, the research team developed a method to identify and prioritize locations with quantitative risk factors of pedestrian and bicycle crashes, not merely based on crash histories. A database of 188 segments and 184 intersections, including detailed geometric and operational elements as well as broad descriptors of the built environment, is assembled for analysis. Logistic regression models for the crash occurrence (crash or not) were developed and a method was developed to create a risk-scoring tool for pedestrians and bicycles at intersections and segments (a total of four scoring tools) using the model results. To demonstrate the application of the risk-scoring tool, the tool was applied to safety projects that were recommended in 2015 All Roads Transportation Safety (ARTS) project lists.

The primary challenge to quantifying the risk for pedestrians and cyclists is the missing measures of exposure and the relatively few pedestrian and bicycle crashes observed on most segments and intersections. The inclusion of the bicycle STRAVA data significantly improved the bicycle models, though the data's ability to represent all bicycle travel accurately is still somewhat uncertain.

In this research, the value of each risk score was derived from the modeling output. All models suffer from the limitations of the input data set. With a larger or different sample for modeling, there is a possibility that the risk scores would be different. The inclusion of exposure measures would likely improve the modeling results. Finally, it is clear that a larger dataset, perhaps derived from GIS or automated data mining tools, would produce a very robust database for a similar modeling effort. Rather than a statewide focus, a regional or Metropolitan Planning Orginization (MPO)-level analysis would likely yield good results leveraging the more detailed spatial data available.

Acknowledgments

This work was funded by the SPR research program of the Oregon Department of Transportation, project SPR 779.

Table 7. Application of Risk Scores to Selected Intersection Projects

Type	City	Project name	Proposed project elements	B/C	Risk score	Risk percentile
Pedestrian	Keizer	RIVER RD NE @ SAM ORCUTT WAY NE	Install no pedestrian phase feature with flashing yellow arrow	17.16	67	>75th
	Beaverton	SW HALL BLVD @ SW NIMBUS AVE	Install pedestrian countdown timer(s)	17	63	>75th
	Eugene	I 105 @ MP 1.8: COBURG RD @ MLK JR BLVD	Install no pedestrian phase feature with flashing yellow arrow; install urban green bike lanes at conflict points	9.87	58	>75th
	Albany	GEARY ST @ QUEEN AVE	Install pedestrian countdown timer(s)	7.13	46	=50th
	Salem	BROADWAY ST NE @ PINE ST NE	Install no pedestrian phase feature with flashing yellow arrow	2.45	53	=75th
Bicycles	Salem	D ST NE @ LANCASTER DR NE	Install urban green bike lanes at conflict points; add no pedestrian phase feature with flashing yellow arrow	20.66	40	<50th
	Salem	FAIRVIEW AVE SE @ 12TH ST SE	Install urban green bike lanes at conflict points	20.06	18	<25th
	Portland	LOMBARD ST @ N INTERSTATE AVE (US 30B)	Install pedestrian countdown timer(s)	16.80	28	<50th
	Albany	GEARY ST @ QUEEN AVE	Install no pedestrian phase feature with flashing yellow arrow; install urban green bike lanes at conflict points	7.13	34	>50th
	Eugene	RIVER RD @ IRVING RD	Install urban green bike lanes at conflict points	2.45	18	<25th

Author Contributions

The authors confirm contributions to the paper as follows: study conception and design: C. Monsere and H. Wang; data collection: C. Chen and Y. Wang; analysis and interpretation of results: Y. Wang, C. Chen, C. Monsere and H. Wang; draft manuscript preparation: all authors. All authors reviewed the results and approved the final version of the manuscript.

References

1. *Highway Safety Manual.* AASHTO, Washington, D.C., 2010.
2. Dill, J. Bicycling for Transportation and Health: The Role of Infrastructure. *Journal of Public Health Policy*, Vol. 30, No. 1, 2009, pp. 95–110. http://dx.doi.org/10.1057/jphp.2008.56
3. Figliozzi, M., P. Johnson, C. Monsere, and K. Nordback. Methodology to Characterize Ideal Short-Term Counting Conditions and Improve AADT Estimation Accuracy Using a Regression-Based Correcting Function. *Journal of Transportation Engineering*, 140, No. 5, 2014, pp. 1–16. http://dx.doi.org/10.1061/(ASCE)TE.1943-5436.0000663
4. Oregon Traffic Crash Summary. *Transportation Data Section, Crash Analysis and Reporting Unit.* Oregon Department of Transportation, Salem, 2017.
5. *Oregon DOT Pedestrian and Bicycle Safety Implementation Plan.* Kittelson & Associates, Inc., Portland, OR, 2014.
6. Berdica, K. An Introduction to Road Vulnerability: What Has Been Done, Is Done and Should Be Done. *Transport Policy*, Vol. 9, No. 2, 2002. pp. 117–127. https://doi.org/10.1016/S0967-070X(02)00011-2.
7. Preston, H., R. Storm, J. D. Bennett, and B. Wemple. *Systemic Safety Project Selection Tool.* Publication No. FHWA-SA-13-019. FHWA, U.S. Department of Transportation, Washington D.C., 2013.
8. Harwood, D. W., D. K. Gilmore, D. J. Torbic, R. Souleyrette, and Z. N. Hans. *United States Road Assessment Program: Pilot Program: Phase II.* Final Report. AAA Foundation for Traffic Safety, Washington D.C., 2008.
9. Carter, D. L., W. W. Hunter, C. V. Zegeer, J. R. Stewart, and H. Huang. *Pedestrian and Bicyclist Intersection Safety Indices.* Publication FHWA-HRT-06-125. FHWA, U.S. Department of Transportation, Washington D.C., 2006.
10. Lagerwey, P. A., M. J. Hintze, J. B. Elliott, J. L. Toole, and R. J. Schneider. *NCHRP Report 803: Pedestrian and Bicycle Transportation Along Existing Roads—ActiveTrans Priority Tool Guidebook.* Transportation Research Board of the National Academies, Washington D.C., 2015.
11. Furth, P. G., M. C. Mekuria, and H. Nixon. Network Connectivity for Low-Stress Bicycling. *Transportation Research Record: Journal of the Transportation Research Board*, 2016. 2587: 41–49.
12. Wang, H., M. Palm, C. Chen, R. Vogt, and Y. Wang. Does Bicycle Network Level of Traffic Stress (LTS)

Explain Bicycle Travel Behavior? Mixed Results from an Oregon Case Study. *Journal of Transportation Geography*, Vol. 57, 2016. pp. 8–18. http://dx.doi.org/10.1016/j.jtrangeo.2016.08.016
13. Eluru, N., C. R. Bhat, and D. A. Hensher. A Mixed Generalized Ordered Response Model for Examining Pedestrian and Bicyclist Injury Severity Level in Traffic Crashes. *Accident Analysis & Prevention*, Vol. 40, No. 3, 2008. pp. 1033–1054. https://doi.org/10.1016/j.aap.2007.11.010
14. Kim, J. K., S. Kim, G. F. Ulfarsson, and L. A. Porrello. Bicyclist Injury Severities in Bicycle–Motor Vehicle Accidents. *Accident Analysis & Prevention*, Vol. 39, No. 2, 2007, pp. 238–251. https://doi.org/10.1016/j.aap.2006.07.002
15. Pai, C. W. Overtaking, Rear-end, and Door Crashes Involving Bicycles: An Empirical Investigation. *Accident Analysis & Prevention*, Vol. 43, No. 3, 2011, pp. 1228–1235. https://doi.org/10.1016/j.aap.2011.01.004
16. Boufous, S., L. de Rome, T. Senserrick, and R. Ivers. Risk Factors for Severe Injury in Cyclists Involved in Traffic Crashes in Victoria, Australia. *Accident Analysis & Prevention*, Vol. 49, 2012, pp. 404–409. https://doi.org/10.1016/j.aap.2012.03.011
17. Schepers, P., and B. den Brinker. What Do Cyclists Need To See To Avoid Single-Bicycle Crashes? *Ergonomics*, Vol. 54, No. 4, 2011, pp. 315–327. https://doi.org/10.1080/00140139.2011.558633
18. Parkin, J., M. Wardman, and M. Page. Models of Perceived Cycling Risk and Route Acceptability. *Accident Analysis & Prevention*, Vol. 39, No. 2, 2007, pp. 364–371. https://doi.org/10.1016/j.aap.2006.08.007
19. Lenguerrand, E., J. L. Martin, and B. Laumon, Modelling the Hierarchical Structure of Road Crash Data—Application to Severity Analysis. *Accident Analysis & Prevention*, Vol. 38, No. 1, 2006, pp. 43–53. https://doi.org/10.1016/j.aap.2005.06.021
20. Jang, K., S. Park, S. Kang, K. Song, S. Kang, and S. Chung. Evaluation of Pedestrian Safety: Pedestrian Crash Hot Spots and Risk Factors for Injury Severity. *Transportation Research Record: Journal of the Transportation Research Board*, 2013. 2393: 104–116.
21. Walker, S. H., and D. B. Duncan. Estimation of the Probability of an Event as a Function of Several Independent Variables. *Biometrika*, Vol. 54, No. 1–2, 1967, pp. 167–179.
22. Washington, S., M. G. Karlaftis, and F. L. Mannering, *Statistical and Econometric Methods for Transportation Data Analysis*, 2nd ed. Chapman & Hall/CRC, Boca Raton, Fla., 2011.
23. Long, S. J. Regression Models for Categorical and Limited Dependent Variables. *Advanced Quantitative Techniques in the Social Sciences*, Vol. 7. 1997.
24. Al-Ghamdi, A. S. Using Logistic Regression To Estimate the Influence of Accident Factors on Accident Severity. *Accident Analysis & Prevention*, Vol. 34, No. 6, 2002. pp. 729–741. https://doi.org/10.1016/S0001-4575(01)00073-2
25. Monsere, C., H. Wang, Y. Wang, and C. Chen. *Risk Factors for Pedestrian and Bicycle Crashes*. Publication FHWA-OR-RD-17-13. Oregon DOT. Salem, 2017.
26. Montgomery, D. C., E. A. Peck, and C. G. Vining. *Introduction to Linear Regression Analysis*. John Wiley & Sons, Hoboken, NJ, 2015.

The Standing Committee on Transportation Safety Management (ANB10) peer-reviewed this paper (18-00738).

Article

TRR
JOURNAL OF THE TRANSPORTATION RESEARCH BOARD

Getting to Zero Deaths on Ohio's Low-Volume Roads

Transportation Research Record
2018, Vol. 2672(32) 40–48

DOI: 10.1177/0361198118782022
journals.sagepub.com/home/trr
SAGE

Victoria F. Beale[1], Derek Troyer[2], Alejandro Chock[3], Cory Hopwood[2], and Mike McNeill[2]

Abstract

Ohio faces the challenge, as do many other states, of how to utilize Highway Safety Improvement Program (HSIP) funding to improve safety on its low-volume roadways while meeting the data-driven safety funding requirements of the Fixing America's Surface Transportation (FAST) Act. Low-volume roads present unique challenges because data is rarely available and other factors, such as roadway ownership, affect the implementation of safety countermeasures on this system. Beginning in 2014, the Ohio Department of Transportation (ODOT) created a township safety signage grant program designed to address the issues with utilizing HSIP funding on low-volume roadways. The grant program's goal was to drive down the number of fatalities, serious injuries, and overall crashes occurring on Ohio's low-volume roads. ODOT took into consideration overriding issues regarding low-volume roads in how it structured the grant program. ODOT also utilized the direction from its Strategic Highway Safety Plan in choosing a safety countermeasure which met the needs of its roadway departure and intersection crash trends. The program has been actively engaged in by Ohio townships and now has a large enough amount of post-safety countermeasure installation data available to quantify its initial success. This paper presents the successful results by highlighting the human capital and comprehensive societal benefit/cost analyses for the first 24 townships with 12 months of post-grant completion crash data.

Ohio's Safety program's mission is to address safety on all of Ohio's roadways, not just those owned by the Ohio Department of Transportation (ODOT) (*1*). This mission is driven by the Highway Safety Improvement Program (HSIP) legislation from the Fixing America's Surface Transportation (FAST) Act (*2*). Achieving Ohio's mission becomes more challenging the lower a roadway's volume dips, as the FAST Act requires a data-driven approach for selection of safety improvements and, historically, lower volume roads have not had sufficient data collected to engage in this required approach. For Ohio's lowest volume roads, which are typically on the township system comprising 34% of Ohio's total centerline miles, the challenge was selecting a data-driven safety countermeasure low in cost and high in return on investment (ROI). The safety countermeasure selected also needed to focus on Ohio's roadway departure and intersection crashes, as they represent the largest number of fatal and serious injuries in Ohio. ODOT believes it has found an initial answer to this challenge as a starting point for its safety investment on Ohio's low-volume roadway system.

Background

Ohio is a home-rule state (*3*). This means Ohio's 2,300+ counties, cities, villages, and townships have direct responsibility for their roadway systems. This group of local governments, often referred to as Ohio's Local Public Agencies (LPAs), must be engaged and encouraged by ODOT to participate in roadway safety improvements through programs such as federally funded grants offered by ODOT's safety section. Based on ownership by center lane mileage, 84% of Ohio's roadway system is outside of ODOT's direct control (see Figure 1). Engagement and encouragement are key

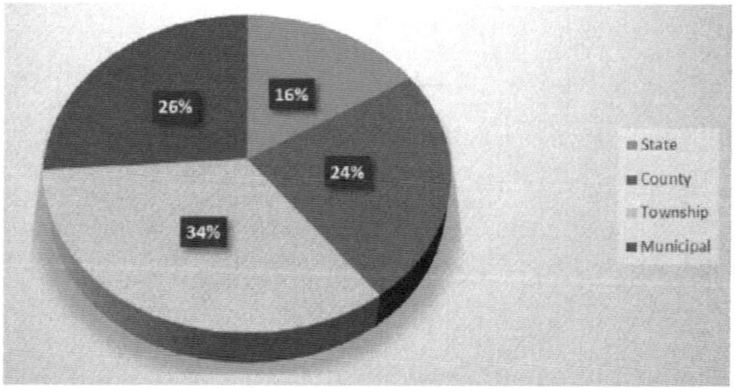

Figure 1. Ohio center lane mileage ownership.

[1]Ohio Local Technical Assistance Program (LTAP) Center, Ohio Department of Transportation, Columbus, Ohio
[2]Ohio Department of Transportation, Columbus, Ohio
[3]Mott MacDonald, Cleveland, Ohio

Corresponding Author:
Address correspondence to Victoria F. Beale: victoria.beale@dot.ohio.gov

Table 1. Summary of Issues to be Addressed in Establishment of Safety Funding Grant Program for Ohio's Townships

Issue number	Issue statement	High-level solution
1	Data-driven approach needed to trigger utilization of federal funds	Use ODOT's existing data warehouse to create locally focused reports to support the data-driven approach
2	Roadway ownership by the LPAs	Select a safety countermeasure directly implementable by the LPA through their normal scope of responsibility and with their existing resources
3	Federal funds carry additional requirements that must be met when utilized	Streamline the grant process so as to avoid triggering the more onerous federal-aid requirements when utilizing federal funding

because ODOT has no direct authority over the extensive LPA roadway system.

Another key to LPA participation in federally funded safety grants is minimizing the impact of federal requirements on the LPAs utilizing federal funding. LPAs in Ohio directly receive a portion of the fuel tax collected to fund their roadway systems. However, this funding level is not sufficient in many jurisdictions to complete basic maintenance for the current system, let alone stretch to implementation of safety countermeasures to increase safety on local roadways. LPAs need the additional federal funding to improve the safety of their roadways, but LPAs do not want to engage in the federal process requirements which accompany acceptance of federal funding.

Given Ohio's environment—low roadway ownership by ODOT and LPA aversion to utilizing federal funds due to the accompanying increased requirements—any program established to help the LPAs, specifically the townships, had to address these two issues in addition to the aforementioned lack of data. These issues, which are discussed next, are also summarized in Table 1.

Issue #1: Getting to Data-Driven—Ohio's GIS Crash Analysis Tool (GCAT)

ODOT has an established system for collecting and mining Ohio crash data. A standardized crash report, the OH-1, is completed by law enforcement for each reported crash in Ohio. This reporting form is utilized by all Ohio law enforcement agencies. The submitted information is collected in a database maintained by the Ohio Department of Public Safety. ODOT transfers this data on a regular basis to their GIS Crash Analysis Tool (GCAT). After significant efforts to locate crashes spatially, GCAT allows ODOT's safety engineers and Strategic Highway Safety Plan partners to report on the crash data through established mapping tools and ad-hoc reporting (see Figure 2).

With the source data of GCAT, ODOT was able to begin addressing the need for a data-driven approach in funding township roadway safety projects. ODOT created township-focused reports to support the types of safety improvements

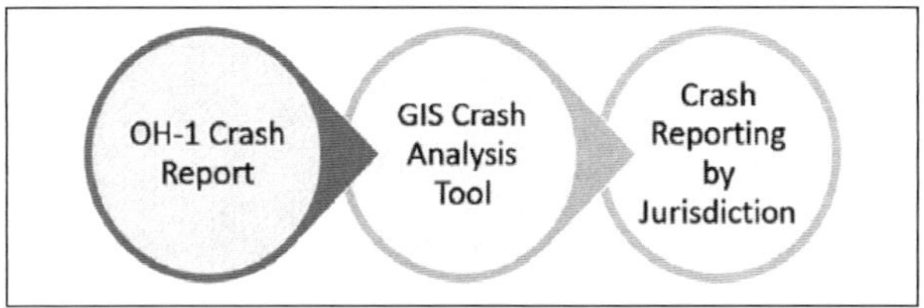

Figure 2. Flow of Ohio's crash data into usable crash reporting for a data-driven crash analysis approach.

implemented on most roadways—hot spot, systemic, and systematic. The reports began with a heat mapping of township-wide crashes (see Figure 3) for the previous five years. This allowed ODOT to determine crash rates and target townships with the greatest need for safety improvements.

Additional tools were created in the form of hot spot maps (see Figure 4) and crash trees (see Figure 5). The maps provided a graphic depiction of a township's roadway system with only crashes occurring on that township's roadways plotted and depicted according to the crash severity. The crash trees provided townships a way to identify systemic roadway crash occurrences, such as a high instance of roadway departures across an entire township system. The combined tools formed the foundation for a data-driven approach to prioritizing award of safety funds to implement proven safety countermeasure(s) on the township roadway system.

Issue #2: Selecting the Best Countermeasure for Townships Given Their Scope of Responsibility and Available Resources

Ohio's townships are often referred to as the 'grassroots' level of government. In Ohio, townships are governed by three elected township trustees and aided by an elected fiscal officer. Engineering responsibility for a township lies, by Ohio law, with the County Engineer for the county where the township is located. Townships traditionally employ a very small roadway staff, who also cross over into other basic township functions. Given these factors, the creation of a

Township	County	District	Population	Mileage	Crashes	Fatalities	Incapacitating	Non-Incapacitating	Possible Injuries
WEST CHESTER	Butler	8	54895	235.82	1137	7	39	181	199
COLERAIN	Hamilton	8	60144	109.95	1024	2	24	70	65
BOARDMAN	Mahoning	4	42518	145.97	985	0	18	152	130
GREEN	Hamilton	8	55660	107.28	891	1	9	75	78
JACKSON	Stark	4	37744	211.48	886	1	16	103	120
PERRY	Stark	4	29167	145.37	846	1	17	109	120
DEERFIELD	Warren	8	25515	109.97	801	0	17	95	139
PLAIN	Stark	4	51997	163.66	765	1	25	88	111
MIAMI	Montgomery	7	45593	91.85	630	1	11	69	107
SPRINGFIELD	Hamilton	8	37587	87.83	600	0	5	36	57
LAKE	Stark	4	25892	148.33	553	1	26	108	82
AUSTINTOWN	Mahoning	4	38001	118.98	552	0	12	87	65
WASHINGTON	Montgomery	7	52991	150.15	535	1	12	98	144
LIBERTY	Butler	8	22819	139.94	501	2	12	59	74
ORANGE	Delaware	6	12464	93.54	494	1	8	68	70

Figure 3. Example of the system-wide crashes heat map data for Ohio's township roadway system.
Note: Data source is the Ohio Department of Transportation.

safety program for townships, utilizing federal funds, had to factor in the limited time and resources of a township's governmental structure.

Utilizing the available crash data, and taking into account the lean nature of township government resources in Ohio, the next challenge for ODOT was to select a countermeasure which would address the crash trends identified in the data and be easy to administer by townships. After reviewing available options which would address the identified crash types and systemic issues, the ODOT Safety Program selected enhanced signage as the targeted countermeasure for the township safety signage grant program. Enhanced signage is the installation of additional safety signage and the replacement of current safety signage so as to be larger and more retroreflective (see Figure 6).

This countermeasure is one of the lowest cost available for increasing safety on a roadway. As the township system accounts for the largest part of Ohio's center lane miles, the low-cost nature of this countermeasure allowed for a far-reaching application across the identified system. Another benefit of selecting this countermeasure was being able to structure the grant program so as to avoid the largest area of concern for an LPA in administering a federally funded safety project, federal regulatory compliance, as discussed next.

Issue #3: Streamlining the Federal-Aid Process for Townships

The federal-aid process allows for flexibility by only requiring compliance with federal law if a condition is a part of a federally funded project. An example of this is triggering of right-of-way requirements if a project is going to acquire land as a part of a federally funded project. If a project does not acquire land, then it can be categorically excluded from right-of-way requirements and move through the federal-aid approval process quicker.

In developing the parameters for funding the selected enhanced signage safety countermeasure on the township system, ODOT's Office of Local Programs assisted the safety section. The Local Programs Office administers all agreements between ODOT and the LPAs statewide for federally funded LPA local-let projects. The Local Programs staff process mapped the proposed safety grant program and identified the potential road blocks to expediting the federal-aid process. The road blocks identified were right-of-way acquisition, environmental clearance, and time and labor tracking. The grant program was then structured to avoid these potential pitfalls through the following grantee requirements:

1) Right-of-Way: All signage provided under the grant was to be installed in existing township right-of-way. No right-of-way could be acquired as a part of the safety project.
2) Environmental: All signage provided under the grant was to be installed as per the following Categorical Exclusion from ODOT's agreement with FHWA:

   ...(4) Activities included in the State's "highway safety plan" under 23 U.S.C. 402 (*4*).

3) Time and Labor Tracking: Only materials would be covered under the grant. No time or labor would be paid for utilizing grant funding. The township grantee would need to install all signs utilizing existing township forces, or with other assistance paid for by the township. No installation costs would be covered under the grant.

One additional road block was also identified as a part of this process and bears noting, even though it did not relate directly to

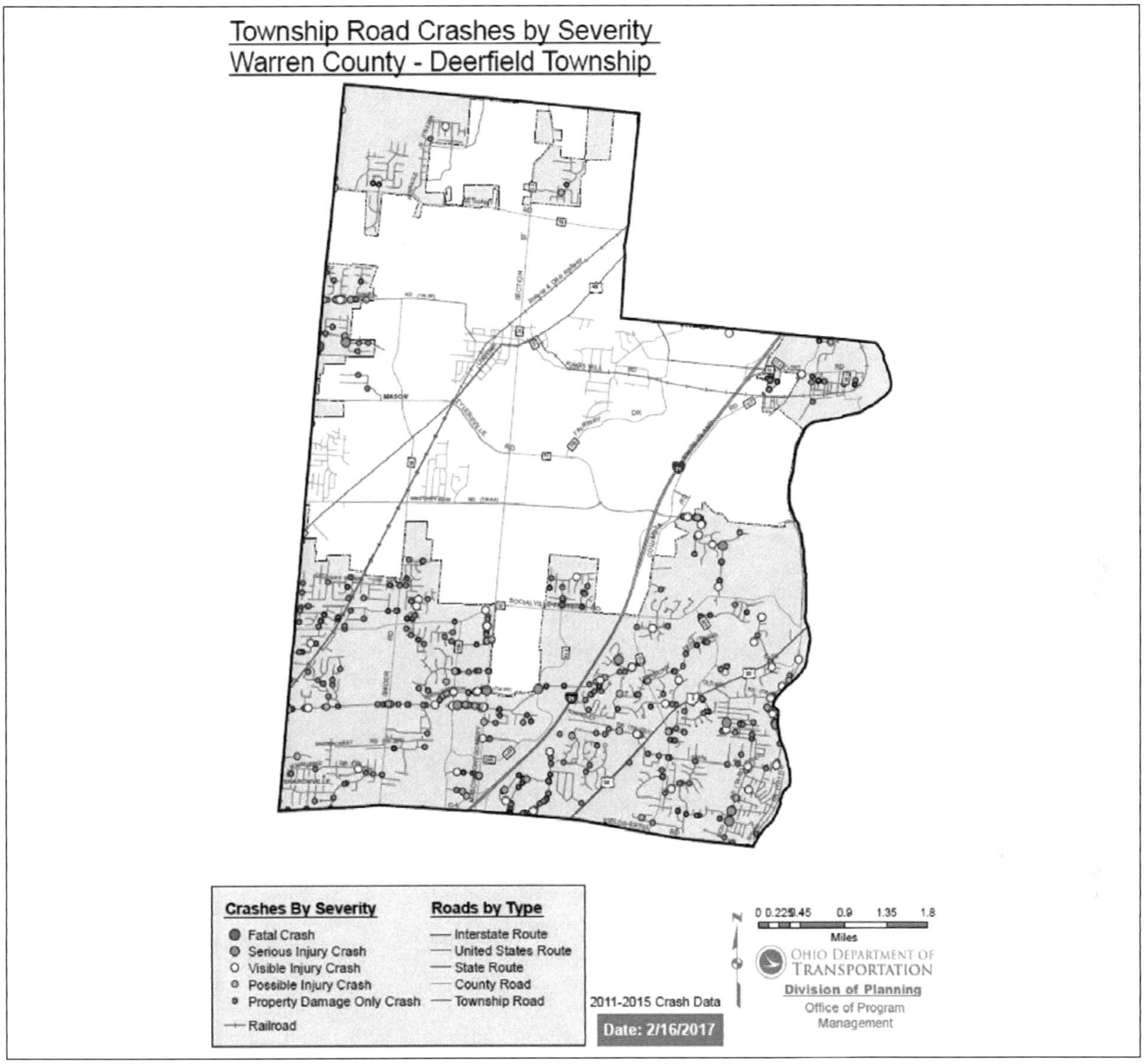

Figure 4. Example of township crash / hot spot map.

the federal-aid requirements. The Local Programs staff felt the ability of a township, with limited resources, to purchase materials directly and then seek reimbursement from ODOT would severely stretch township budgets. In light of this issue, the grant was structured to require LPAs to allow ODOT to make payments for the material purchases directly to the vendors on behalf of the townships. This eliminated the need for a township to have additional cash flow in its budget to purchase materials and then wait for reimbursement.

How the Grant Process Works

The Ohio township safety signage grant program began in state fiscal year 2014, and at the time this paper is being written the program is entering its fifth year. Details of how the program process works from a high-level perspective are discussed in this section to demonstrate the interaction between the grantor, ODOT, and the grantees, the selected Ohio townships invited to apply for the grant funding. Following this section on how the grant process works is the highlight of this paper—the initial data demonstrating the success of this program for Ohio's low-volume roads.

Step 1: Identifying Potential Grantees

In the fall of each year, the ODOT Safety section finalizes crash data for the previous year and creates a new heat map of township-wide crashes (see Figure 3). The heat map

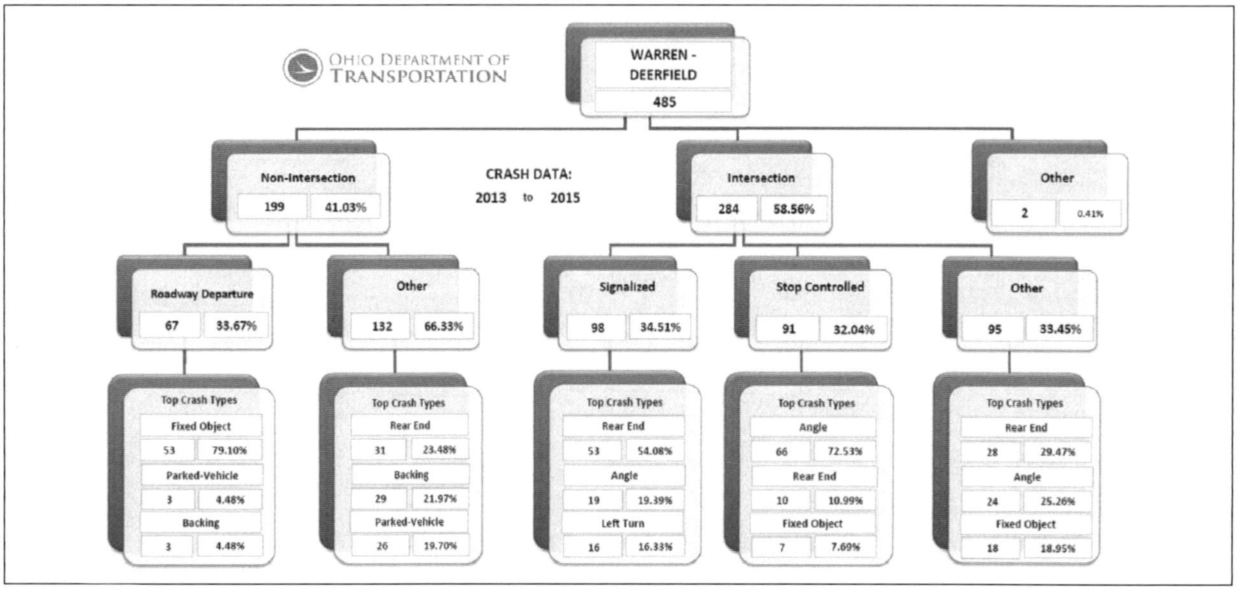

Figure 5. Example of a township crash tree.

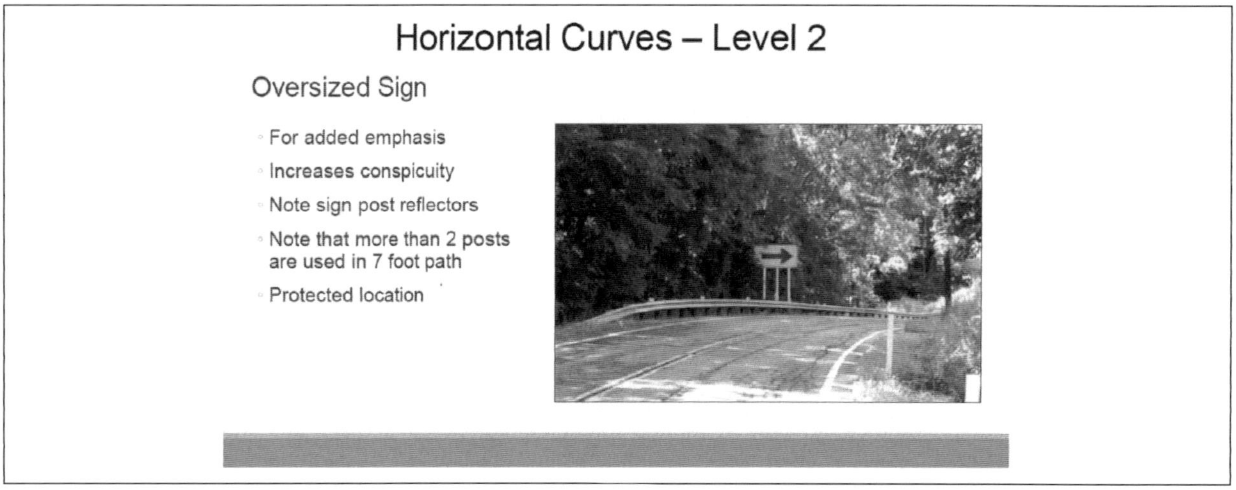

Figure 6. Example of guidance for oversized sign installation on township roadways.

allows townships who rank at the top to be targeted for invitation to apply for the township safety signage grant program. Targeted townships must meet two criteria:

1) Have a greater than average crash rating across their entire township based on their township's crash history over the previous five years and
2) Not have previously received funding under the grant program.

The top 100 townships meeting these criteria each year are invited to apply for up to $50,000 per township in posts, signs, and hardware to improve the safety of their roadways. The materials are covered at 100% utilizing federal HSIP funds. Potential grantees are invited to apply and are made aware of a pre-grant meeting which their township must attend to apply for the grant. The invitation includes a packet of information for the potential grantee to begin working on their application:

1) The township's crash data in the form of their crash map and crash tree
2) A copy of the sign grant application
3) A copy of the information which will be reviewed at the required pre-grant meeting
4) Instructions on how to register for the pre-grant meeting

Step 2: The Pre-Grant Meeting

At the pre-grant meeting, townships learn the grant process steps. Townships are also provided background on how the crash data were developed for their township and are given an opportunity to ask specific questions about their crash data. Townships learn how to utilize their safety data to first address crash hot spots on their systems, followed by a systemic approach to mitigate same or similar locations on their roadways and then encouraged to utilize any funds remaining within their $50,000 grant limit to systematically address safety on their township roads. The systemic and systematic approaches are guided by the township crash trees.

From a technical assistance perspective, the meeting also includes an extensive presentation on safety sign package templates townships can choose to adopt for installation at their intersection and curve crash locations. The safety sign package presentation reviews in depth the requirements for installation of the various safety signs offered under the grant, including a discussion of relevant Ohio Manual of Uniform Traffic Control Devices (OMUTCD) sections.

Upon leaving the pre-grant meeting, township grant applicants have a plan on how to determine their safety signage needs based on their crash data. They compile their sign, post, and hardware needs utilizing this plan and submit the needs on their grant application.

Step 3: Post-Grant Award Processing

Following award of the grants, ODOT proceeds with notifying the grantees of their awards, issuing an LPA agreement for signature by the grantee, and confirming the signs, posts, and hardware orders submitted are accurate. Once the agreement is signed by the LPA and the orders are confirmed, ODOT facilitates internal processes including right-of-way and environmental clearance leading up to a federal authorization request to FHWA to fund the project. Upon receipt of federal authorization, ODOT issues purchase orders for the requested materials and completes the purchasing process on behalf of the grantees. On average, this post-grant award processing can take three to six months, depending on a number of factors including state fiscal year end and product availability from the vendors.

Step 4: Sign Installation

Installation of the signs, posts, and hardware is the responsibility of the grantee. Grantees are initially provided a year after the federal authorization date to complete installation of their grant materials; however, grantees are also afforded the opportunity to request an additional year to complete installation upon good cause. Once installation is completed, the grantee submits a certification verifying all materials provided under the grant have been installed.

Step 5: Project Closeout

The installation certification form submitted by the grantee triggers a final project closeout at ODOT. Upon receipt of the form, ODOT works through its Local Programs Office to process the federally authorized project for closeout with FHWA and request reimbursement of ODOT's expended funds from the federal HSIP funds.

As a quality assurance measurement, ODOT also then randomly selects grantee townships who have participated in the program for a review of their signage installation. The review verifies the materials were installed in compliance with the OMUTCD requirements. Any township found to have incorrectly installed their materials is provided a detailed description of the installation issues and afforded time to correct their installations. A follow-up review is conducted, either once the township communicates the issues have been resolved, or at 90 days after the original notification is issued.

Success as Reported by the Numbers

The data-driven approach looks to data after implementation to measure success. ODOT waited until 24 of the originally invited townships had at least 12 months of post-installation crash data before engaging in a comparison to measure success. The comparison measured the decrease in crashes in three categories—overall crashes, serious injuries, and fatalities. The period measured for each township was the 12 months of calendar year 2016.

Overall crashes for the 24 townships decreased 10%. This percentage was derived by looking at the average number of crashes per year for the three years immediately preceding sign installation against the 12-month period following sign installation. Overall crashes for the identified townships were 1433.33 per year prior to installation and 1294 per year for the same townships after installation.

Serious injury crashes for the 24 townships decreased 35%. This was measured as an annual average for the identified townships of 39 serious injury crashes prior to signage installation and 25 serious injury crashes for the same townships in the 12 months following signage installation.

Fatal injury crashes for the 24 townships decreased 100%. This was measured as an annual average for the identified townships of 4.67 fatalities prior to signage installation and 0 fatalities for the same townships in the 12 months following signage installation.

The Economic Impact of the Crash Reductions

So how do the demonstrated crash reductions translate to benefit/cost analysis ratio or a ROI for the implemented safety countermeasure? The ODOT Safety section utilized the pre- and post-installation data to perform both a human capital cost analysis and a comprehensive societal cost

Table 2. Grant Amounts Spent on Behalf of the Townships for the 24 Townships with 12 Months of Post-Installation Crash Data

#	PID	Township	County	Cost
1	96230	Lake	WOO	$ 44,320.61
2	96154	Jackson	SEN	$ 21,900.56
3	96114	Canton	STA	$ 49,320.21
4	98440	Ravenna	POR	$ 7,046.11
5	96155	Lake	STA	$ 24,062.36
6	98457	Urbana	CHP	$ 7,563.00
7	96170	Violet	FAI	$ 48,239.34
8	98410	Belpre	WAS	$ 24,042.94
9	98437	Painesville	LAK	$ 11,740.29
10	98435	Montville	MED	$ 11,993.34
11	98291	Shortcreek	HAS	$ 733.86
12	98432	Medina	MED	$ 15,788.30
13	96160	Milton	WAY	$ 8,658.29
14	96161	Nimishillen	STA	$ 22,645.28
15	96162	Perry	STA	$ 36,427.74
16	96201	Madison	COL	$ 4,012.72
17	97522	Green	HAM	$ 4,703.84
18	97533	Washington	MOT	$ 21,465.22
19	98419	Etna	LIC	$ 25,482.42
20	98428	Madison	LAK	$ 25,754.37
21	100221	Allen	HAN	$ 14,689.39
22	98438	Porter	SCI	$ 1,903.10
23	97528	Perkins	ERI	$ 51,096.29
24	98451	Troy	RIC	$ 39,334.71
		Total Spent		$ 522,924.29

Note: PID = Project Identification Number. Perkins Township in Erie County's award exceeded the standard $50,000 grant limit as a result of the sign materials contract pricing changing mid-grant and the materials being ordered costing more than originally anticipated by ODOT. The program during the first four years provided $1,000,000 in funding for township grantees. Based on the demonstrated success which will be discussed later, the level of funding has been raised to $2,000,000 for the fifth year of the program and the number of townships invited to apply for the grant has been doubled.

analysis. They first compiled the amounts spent for materials for each of the 24 townships, as displayed in Table 2.

The crashes were then categorized into the pre-established crash types per the Highway Safety Manual (HSM)—fatal (K), disabling injury (A), evident injury (B), possible injury (C), and property damage only (O). Calculations were performed to arrive at the human capital crash cost and the comprehensive societal crash cost for each crash type, calibrating the amounts based on the crashes occurring in Ohio and during the period of time measured. Using the calculated cost per crash type, the total before and after crash costs were calculated for the human capital measurement and the comprehensive societal crash measurement (see Table 3). The difference in the costs from the pre- and post-installation periods were then divided by the total grant costs to arrive at a 21.47:1 benefit/cost, or ROI, for the human capital measurement approach (see Table 4) and a 62.59:1 benefit/cost, or ROI, for the comprehensive societal crash measurement approach (see Table 5).

Table 3. Calculated Human Capital and Comprehensive Societal Crash Costs in Ohio for the Measured Post-Installation Time Period

Crash severity	Comprehensive societal cost	Human capital cost
K	$ 5,731,155	$ 1,669,104
A	$ 303,038	$ 149,276
B	$ 110,683	$ 56,146
C	$ 62,311	$ 38,056
O	$ 10,046	$ 8,576

Note: *Highway Safety Manual*. AASHTO, Washington, D.C., 2010. Comprehensive Crash Cost = The cost of each crash by either crash type or severity that includes the human capital costs in addition to non-monetary costs related to the reduction in the quality of life in order to capture a more accurate level of the burden of injury. Human Capital Crash Costs = The cost of each crash by either crash type or severity that includes monetary losses associated with medical care, emergency services, property damage, and lost productivity.

ODOT's Goal to Justify a Countermeasure's Implementation

In general, ODOT seeks to have countermeasures which exceed a benefit/cost ratio of 1:1. This is the level sought as detailed in ODOT's Safety Study Guidelines (*5*). With the safety countermeasure of enhanced signage for better visibility on Ohio's low-volume, township roadway system, the human capital costs analysis and the comprehensive societal costs analysis both far exceed this goal.

A Solid Safety Countermeasure Choice for Ohio

Based on the initial, post-sign installation data from the township safety signage grant program, ODOT has a great, data-driven starting point for low-cost safety improvements on Ohio's low-volume roadways. A benefit/cost or ROI utilizing the most conservative calculations still provides a ratio of 21.47:1, far exceeding the standard 1:1 ratio sought to justify most roadway safety projects in Ohio. The benefits will continue for many years to come on these township roads.

As more townships apply and utilize the HSIP-funded safety signage grant program, it is anticipated that the overall benefit/cost ratio will decrease. This is because of the initial targeting of townships with more severe (hot spot) and pervasive (systemic/systematic) crash problems; however, projections support the benefit/cost ratio remaining well above the 1:1 requirement to fund a safety project in Ohio. The costs associated with the signage, post, and hardware purchases under the grant are minimal compared with other

Table 4. Human Capital Costs Analysis to Arrive at Cost/Benefit Ratio (ROI)

Human capital costs						
Severity	Before	Cost	After	Cost	Delta	Delta cost
K	4.67	$7,789,152.00	0	$0.00	−4.67	-$7,789,152.00
A	39.00	$5,821,764.00	25	$3,731,900.00	−14.00	-$2,089,864.00
B	186.67	$10,480,586.67	186	$10,443,156.00	−0.67	-$37,430.67
C	216.33	$8,232,781.33	209	$7,953,704.00	−7.33	-$279,077.33
O	1123.33	$9,633,706.67	1003	$8,601,728.00	−120.33	-$1,031,978.67
Totals		$41,957,990.67		$30,730,488.00		-$11,227,502.67
				Township average		$21,788.51
				Total spent		$522,924.29
				Benefit/cost		21.47

Table 5. Comprehensive Societal Costs Analysis to Arrive at Cost/Benefit Ratio (ROI)

Comprehensive societal costs						
Severity	Before	Cost	After	Cost	Delta	Delta cost
K	4.67	$26,745,390.00	0	$0.00	−4.67	-$26,745,390.00
A	39.00	$11,818,482.00	25	$7,575,950.00	−14.00	-$4,242,532.00
B	186.67	$20,660,826.67	186	$20,587,038.00	−0.67	-$73,788.67
C	216.33	$13,479,946.33	209	$13,022,999.00	−7.33	-$456,947.33
O	1123.33	$11,285,006.67	1003	$10,076,138.00	−120.33	-$1,208,868.67
Totals		$83,989,651.67		$51,262,125.00		-$32,727,526.67
				Township average		$21,788.51
				Total spent		$522,924.29
				Benefit/cost		62.59

types of safety improvements. These lower costs and the ability to implement the countermeasure on the largest percentage of Ohio's roadway system will factor into this program being successful for years to come.

Lessons Learned

Any program goes through growing pains when first established and the township safety sign grant program was no exception. Fortunately many of the lessons learned were easy remedies. During the first year, the program awards were made utilizing the materials contract pricing available at that time and not factoring in the federal authorization processing time needed before purchase orders could be issued. Once the projects received federal authorization, the materials contracts had been reissued with different pricing, which caused amendments to the project agreements and the federal authorization amounts. Moving forward, ODOT adjusted the timing of the awards to coincide better with the materials contract award periods so amounts would not have to be adjusted between award and actual issuance of the purchase orders.

Another lesson learned was the utilization of electronic agreement processing between the grantor and grantee, versus executing a paper agreement. During the first two years, ODOT issued paper agreements for signature by the grantees. This meant mailing the agreements to the townships and waiting for the signed originals to be mailed back. Beginning with year three, ODOT began issuing the agreements electronically to the townships for signature. The electronic agreement processing cut the time down from three to six weeks on average to mail out and receive the agreement back to between one to three days. This time savings allows ODOT to get the materials into the hands of the townships quicker for installation.

ODOT continually looks for improvement opportunities in all of its processes. Efficiencies will continue to be applied to the program as they are discovered and tested for effectiveness.

Not Resting on Our Laurels

While ODOT's engineers are pleased to have developed a data-driven, successful township safety signage grant program for implementing a safety countermeasure on Ohio's low-volume roads, it is important to remember everything in time will run its course. Eventually every Ohio township could hypothetically be invited to participate in the grant program as more townships receive funding, improve their roadways, and move off the list. There are 1,308 townships in Ohio, and to date with the modification of inviting 200 townships in year five, almost half of them have now been invited to apply for grants. In addition, a modified version of this program is being concurrently utilized on Ohio's county roadways to assist with signage enhancements. ODOT's engineers must and will continue to look for new countermeasures to enhance and augment this current safety initiative for Ohio's low-volume roads. Getting to zero fatalities will take a continued combination of strong partnerships and utilization of federal funding flexibilities, along with the right countermeasures, to reach the only acceptable number of deaths on our roadways—zero.

References

1. Ohio Strategic Highway Safety Plan, Local Road Safety. *Ohio Department of Transportation, Office of Systems Planning, Safety Section.* https://www.dot.state.oh.us/Divisions/Planning/ProgramManagement/HighwaySafety/SHSP/Documents/Ohio_SHSP_Plan.pdf. Accessed July 26, 2017.
2. FAST Act § 1113; 23 U.S.C. 148, 2015. https://www.gpo.gov/fdsys/pkg/USCODE-2011-title23/pdf/USCODE-2011-title23-chap1-sec148.pdf. Accessed June 4, 2018.
3. Hartmann, P. *Ohio's Constitution Gives Municipalities "Home Rule" Self-Government Authority.* Ohio State Bar Association. https://www.ohiobar.org/ForPublic/Resources/LawYouCanUse/Pages/Ohio%27s-Constitution-Gives-Municipalities-Home-Rule-Self-Government-Authority.aspx. Accessed July 11, 2017.
4. ODOT NEPA Assignment Categorical Exclusion Guidance. *Ohio Department of Transportation, Office of Environmental Services.* https://www.dot.state.oh.us/Divisions/Planning/Environment/NEPA_policy_issues/EnvironmentalDocumentation/Documents/CE_Guidance_NEPA.pdf. Accessed July 28, 2017.
5. Safety Studies Guidelines. *Ohio Department of Transportation, Office of Systems Planning, Safety Section.* http://www.dot.state.oh.us/Divisions/Planning/ProgramManagement/HighwaySafety/HSIP/Safety_Study/Safety%20Study%20Guidelines.pdf. Accessed July 12, 2017.

The Standing Committee on Transportation Safety Management (ANB10) peer-reviewed this paper (18-01284).

Article

TRR
JOURNAL OF THE TRANSPORTATION RESEARCH BOARD

Transportation Research Record
2018, Vol. 2672(32) 49–60

DOI: 10.1177/0361198118772704
journals.sagepub.com/home/trr
SAGE

Integrated Modeling Approach for Non-Motorized Mode Trips and Fatal Crashes in the Framework of Transportation Safety Planning

Jaeyoung Lee[1], Mohamed Abdel-Aty[2], Qing Cai[2], Ling Wang[2,3], and Helai Huang[4]

Abstract

In recent decades, considerable efforts have been made to incorporate traffic safety into long-term transportation plans (LTTPs), a process which is often termed transportation safety planning (TSP). Although some researchers have attempted to integrate transportation plans and safety by adopting transportation planning data (e.g., trip generation) for estimating traffic crash frequency at the macroscopic level, no studies have attempted to develop trip and safety models in one structure simultaneously. A Bayesian integrated multivariate modeling approach is suggested for estimating trips and crashes of non-motorized modes (i.e., walking and cycling). American Housing Survey (AHS) data were collected from the U.S. Census Bureau and were used for the proposed approach. In the first part of the proposed model, the probabilities of choosing walking and cycling modes were estimated, and the estimated probabilities were converted to trips by multiplying the number of sampled households. In the second part, the estimated trips were fed into crash prediction models (or safety performance functions) as an exposure variable. The modeling result revealed many contributing factors for pedestrian/bicycle trips and crashes. Possible shared unobserved features between pedestrian and bicycle trips, and between pedestrian and bicycle crashes, were accounted for by adopting a multivariate structure. In addition, it was found that the crash models with the estimated exposures outperform those with the observed exposures. It is expected that the integrated modeling approach for trips and crashes in this study will provide great insights into the future directions of TSP.

Transportation safety planning (TSP) aims at reducing traffic fatalities and injuries by encouraging comprehensive, system-wide, multimodal, data-driven, and proactive transportation planning processes that incorporate traffic safety into long-term transportation plans (LTTPs). The recent federal laws require that the planning processes of state departments of transportation (DOTs) and metropolitan planning organizations (MPOs) be consistent with Strategic Highway Safety Plans (*1*), such as the Moving Ahead for Progress in the 21st Century (MAP-21) Act of 2012 and Fixing America's Surface Transportation (FAST) Act of 2015. In this context of TSP, much effort has been put into integrating planning and safety. Washington et al. conducted research to address traffic safety issues within the LTTPs of state DOTs and MPOs (*2*). The authors developed a tool, PLANSAFE, to forecast safety effects at the zonal level (e.g., traffic analysis zones (TAZs)). Lovegrove and Sayed developed a series of macroscopic crash prediction models to provide a safety planning decision support tool that facilitates a proactive approach to community planning which addresses safety before problems occur (*3*). In a subsequent study, Lovegrove and Sayed discussed the application of the developed macroscopic crash prediction models in safety planning (*4*). Naderan and Shahi developed a series of aggregate crash prediction models that are consistent with the trip generation step of the four-step demand forecasting models (*5*). Similarly, Abdel-Aty et al. investigated the association between crash frequency

[1]Center for Advanced Transportation Systems Simulation (CATSS), Department of Civil, Environmental, and Construction Engineering, University of Central Florida, Orlando, FL
[2]Department of Civil, Environmental, and Construction Engineering, University of Central Florida, Orlando, FL
[3]College of Transportation Engineering, Tongji University, Shanghai, China
[4]Urban Transport Research Center, School of Traffic and Transportation Engineering, Central South University, Changsha, Hunan, China

Corresponding Author:
Address correspondence to Jaeyoung Lee: Jaeyoung@knights.ucf.edu

and various types of trip productions and attractions in combination with roadway features (*6*). Pirdavani et al. developed planning-level zonal crash prediction models (*7*).

Some researchers have explored the best geographic units for macroscopic safety modeling. Among the geographic units, TAZs have been most widely used for TSP as they are the only zone system related to transportation. Nevertheless, TAZs are developed for transportation planning, not for safety analysis. Thus, they are not homogenous in size and the studies based on TAZs are dependent on the selected sample. Lee et al. also pointed out the limitations of TAZs, and developed a new zone system for TSP (*8*). The authors showed that the crash models based on the new zone system (i.e., traffic safety analysis zones) had a better fit compared with TAZ-based models. Abdel-Aty et al. compared TAZs, block groups and census tracts and found that the significance of independent variables is not consistent in the models based on different geographic units (*9*). Xu et al. conducted a sensitivity analysis to investigate the effects of geographic units on modeling results, which is termed the modifiable areal unit problem (*10*). Cai et al. proposed a methodology to compare different zone systems for TSP and showed that the model based on traffic analysis districts (aggregation of TAZs) has the best model performance (*11*). Lee et al. suggested a different approach for TSP (*12*). Instead of exploring crash location, the authors analyzed residence characteristics of at-fault drivers who were responsible for the traffic crash. Dong et al. explored macro-level safety accounting for spatial and temporal autocorrelations and their interactions (*13*). Siddiqui and Watkins evaluated the long-range regional safety of alternative growth scenarios (*14*).

In recent years, traffic safety of non-motorized users (i.e., pedestrians and bicyclists) has been focused on by many researchers (*15–24*). Even though these active modes of transportation have no negative impact on the environment and enhance physical health, safety concerns form the largest impediment to their adoption as a preferred option over motor vehicles for short distance trips. For this reason, many researchers have examined traffic safety of non-motorized users. Wier et al. investigated pedestrian crashes and showed that traffic volume, arterials without transit, proportions of land area zoned for neighborhood commercial and residential neighborhood commercial uses, employee and resident populations, and people living in poverty had a positive relationship with pedestrian crashes (*15*). In contrast, land areas, and the proportion of elderly population had a negative relationship. Siddiqui et al. estimated spatial models for pedestrian and bicycle crashes (*16*). The developed models identified many contributing factors such as roadway with a lower speed, intersection count, household income, dwelling units, population density, car ownership, parking cost, or employment. Lee et al. identified contributing factors for pedestrian crashes using a multivariate model (*17*). The authors revealed that vehicle-miles-traveled, population, proportion of African Americans, proportion of Hispanics, proportion of households without a vehicle, accommodation facility rooms, employment and school enrollment density, traffic signals per mile, and truck percentage have a positive relationship with pedestrian crashes. On the other hand, proportion of roadway with speed limit more than or equal to 55 mph is negatively related to pedestrian crashes.

As mentioned, many research studies have been done for TSP and macroscopic safety analysis of pedestrian and bicycle crashes. Still there is much room for improvement. Some researchers have attempted to integrate transportation planning and traffic safety by using transportation planning data (e.g., trip generation) as explanatory variables for estimating traffic crash frequency at the macroscopic level. In this case, however, it would be difficult to distinguish the factors related to trips (or exposure) from the actual contributing factors for traffic crashes. In the latest study by Shah et al., the authors implemented a two-step process (*22*). The first step was the development of Tobit and generalized linear regression models for predicting pedestrian trips. The second step developed negative binomial and zero-inflated negative binomial models for pedestrian crashes using the predicted trips in the first step. By adopting the two-step process, it was possible to decide whether a factor was relevant to trip generation or crash occurrence. Nevertheless, this study still estimated pedestrian trips and crashes separately. Until now, no study has attempted to develop trip and safety models in one modeling structure. The idea to integrate trip and safety models was motivated by the assumption that walking/cycling trips and their related crashes interact with each other. In other words, walking/cycling trips have an effect on the numbers of pedestrian/bicycle crashes, and vice versa. Integrating the two models would introduce one more constraint for the models estimation, which could lead to a better model performance.

In this study, a Bayesian integrated bivariate modeling approach is suggested for simultaneously estimating trips and crashes of non-motorized modes (i.e., walking and cycling) in one structure. The proposed approach has two model components. The first component is to estimate the probabilities of choosing walking and cycling, and the estimated probabilities are converted to the number of trips. The second component is to estimate the number of crashes using the estimated trips in the first component along with other candidate variables. To account for unobserved shared factors between pedestrian and bicycle trips and crash counts, a bivariate structure is applied. It is expected that the model's performance could be improved by linking trip and crash estimation components. The following sections cover methodology, data preparation, results and discussion, and summary and conclusions. In the methodology section, detailed descriptions of the suggested approach are provided. The collection and processing procedures are briefly explained in the data preparation section. Then, the modeling results and comparison are discussed in the results and discussion section. Lastly, the final section summarizes and concludes the paper.

Methodology

As described previously, the proposed approach has two components. The first component is a logistic model that estimates the probabilities of using pedestrian and bicycle modes. The second component is a Poisson lognormal model that estimates the number of crashes using the expected trips from the first component. The integrated approach has been applied in recent studies. Wang et al. integrated crash frequency and real-time safety models for expressways (*25*). The integrated model connected the crash frequency and the real-time models by adding the log of the expected crash frequency in the real-time model component. The authors concluded that the crash frequency model and real-time risk model support each other and that the integration improved the overall performance.

To aid understanding, the whole model estimation processes for integrated and non-integrated approaches are depicted in Figures 1 and 2, respectively. Equations 1–5 are for the first component. The following Equation 1 shows that the observed number of trips follows a binomial distribution with a probability of using walking or cycling, and the number of sampled households as trials.

$$R_i^m \sim Binomial(p_i^m, \mathrm{HH}_i) \tag{1}$$

where

R_i^m is the observed number of trips of mode m from the sampled households (HH_i) in area i, and m could be either pedestrian or bike,

p_i^m is the expected probability of using mode m in area i, and

HH_i is the number of the sampled households in area i.

The expected probability of walking or cycling is as follows:

$$logit(p_i^m) = \beta_0 + \sum_{k=1}^{K} \beta_k^m X_{ki}^m + \varepsilon_i^m \tag{2}$$

where

p_i^m is the expected probability of using mode m in area i,

β_0 is the intercept,

X_{ki}^m is the kth explanatory variable showing characteristics of area i for mode m,

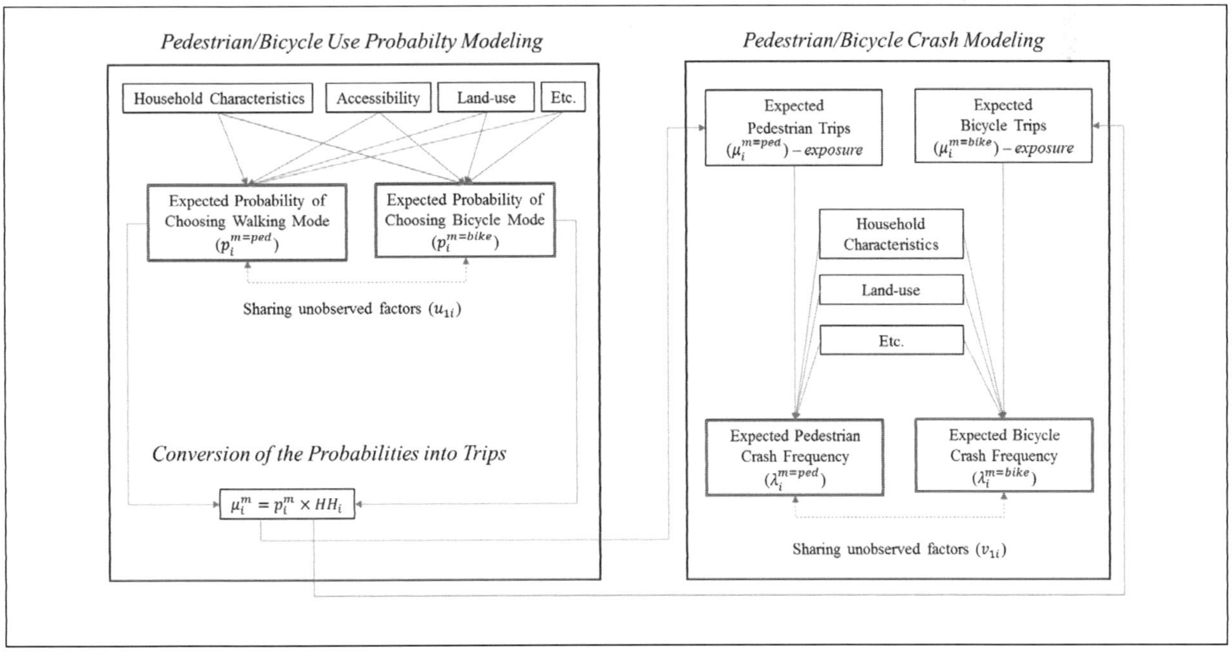

Figure 1. Model estimation process of the integrated approach.

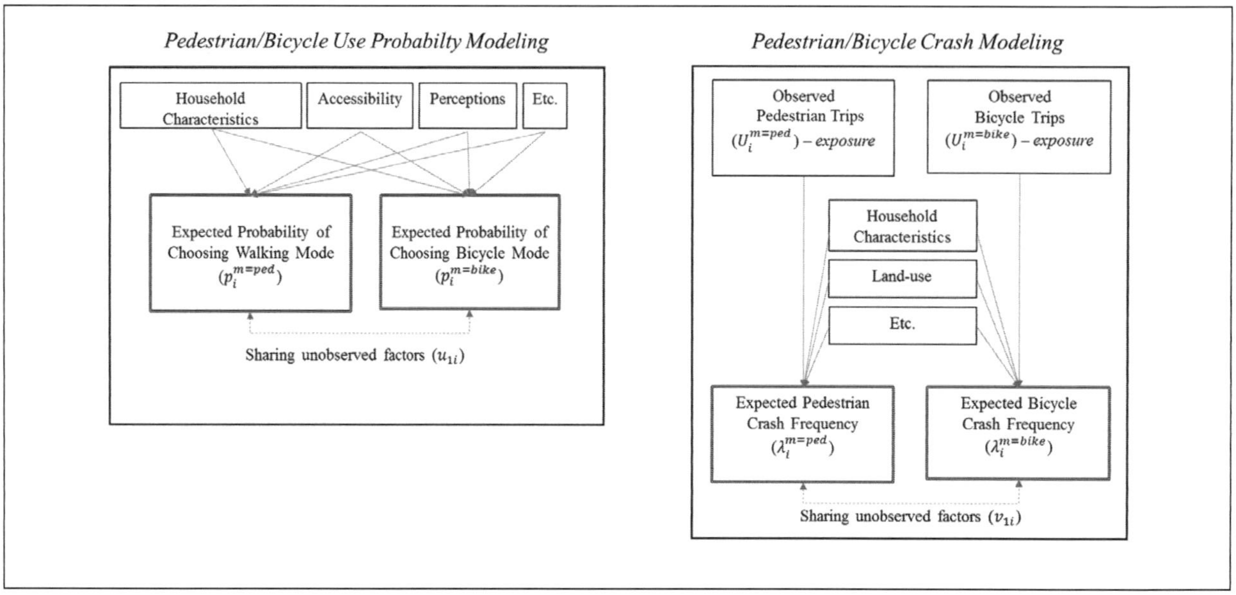

Figure 2. Model estimation process of the non-integrated approach.

β_k^m is the kth coefficient estimate of the kth explanatory variable for mode m, and
ε_i^m is the area-mode specific random error term representing normal heterogeneity of area i for mode m.

To account for possible shared common features between pedestrian and bicycle trips and between pedestrian and bicycle crashes, a multivariate structure was adopted (*26*, *27*).

$$\varepsilon_i^{m\,=\,\text{ped}} = a_1 u_{1i} \tag{3}$$

$$\varepsilon_i^{m\,=\,\text{bike}} = a_2 u_{1i} + a_3 u_{2i} \tag{4}$$

where
ε_i^m is the area-mode specific random error term representing normal heterogeneity of area i for mode m (m = ped is pedestrian and m = bike is bicycle),
both u_{1i} and u_{2i} are the independent random variables, which follow a normal distribution (0, τ) for area i,
τ is the precision parameter that is the inverse of the variance; it follows a gamma distribution (0.1, 0.0001), and
a_1, a_2, and a_3 are the coefficients for the independent variables, u_{1i} and u_{2i}.

Subsequently, the probability of using walking or cycling is converted to the number of trips using

$$\mu_i^m = p_i^m \times \text{HH}_i \tag{5}$$

where
μ_i^m is the estimated number of trips of mode m in area i, and
p_i^m is the expected probability of using mode m in area i.

Equations 6–10 are part of the second component. The observed number of fatal crashes follows a Poisson distribution with the expected number of fatal crashes, λ_i^m.

$$Y_i^m \sim Poisson\left(\lambda_i^m\right) \tag{6}$$

where
Y_i^m is the observed number of fatal crashes involving mode m in area i, which follows a Poisson distribution, and
λ_i^m is the expected number of fatal crashes involving mode m in area i.

The functional form for the integrated approach is shown in Equation 7. The expected number of trips from the first component is used as an exposure variable in the second model component (Figure 1).

$$\log(\lambda_i^m) = \alpha_0 + \alpha_1 \log(\mu_i^m) + \sum_{l=2}^{L} \alpha_l^m X_{li}^m + \theta_i^m \tag{7}$$

where
α_0 is the intercept,
α_1 is the coefficient of μ_i^m,
α_l is the coefficient of X_{li}^m, and
θ_i^m is the area-mode specific random error term representing normal heterogeneity of area i for mode m.

The function for the non-integrated approach is presented in Equation 8. It is different from the integrated approach in that the observed number of crashes is directly used instead of the expected number (Figure 2).

$$\log(\lambda_i^m) = \alpha_0 + \alpha_1 \log(U_i^m) + \sum_{l=2}^{L} \alpha_l^m X_{li}^m + \theta_i^m \quad (8)$$

where U_i^m is the observed number of trips of mode m in area i.

As in the first component, a multivariate structure is implemented to take possible unobserved shared components between pedestrian and bicycle crashes.

$$\theta_i^{m=\text{ped}} = b_1 v_{1i} \quad (9)$$

$$\theta_i^{m=\text{bike}} = b_2 v_{1i} + b_3 v_{2i} \quad (10)$$

The Bayesian inference was employed to estimate the model parameters and their posterior distributions in this study. The models were run considering informative normal priors for β_k^m and α_l^m. The informative priors were obtained from the preliminary non-Bayesian logistic and negative binomial models that were estimated by maximum likelihood estimation (*28*). For a_1, a_2, a_3, b_1, b_2, and b_3, the non-informative normal $(0,10^{-6})$ priors were applied.

Among the advantages of the Bayesian approach are a more natural interpretation of parameter intervals, Bayesian credible interval (BCI) and the freedom of obtaining true parameter density (*29*). In this study, the variables for which 90% BCI of the posterior parameter estimates had the same signs were considered significant . Deviance Information Criterion (DIC) was computed to compare the model performance. The model with the smallest DIC is estimated to be the model that has the best predictability if the model complexity is the same (*30*).

$$DIC = 2 \times \bar{D} - \hat{D} \quad (11)$$

where
$\bar{D}$ is the posterior mean of deviance D,

$\hat{D} = 2 \times (p(y|\bar{\theta}))$, and

$\bar{\theta}$ is the posterior mean of θ

Data Preparation

The American Housing Survey (AHS) data as of 2013 from the U.S. Census Bureau and the Fatality Analysis Reporting System (FARS) data as of 2013 from the National Highway Traffic Safety Administration (NHTSA) were used for the proposed approach. The AHS is the most comprehensive national housing survey in the U.S. It aims to provide a current and continuous series of data on selected housing and demographic characteristics (*31*). The number of the variables provided from the AHS is over 200, including demographic, socioeconomic, walking or cycling environments, public transportation systems, and accessibility to various facilities, in major metropolitan areas. Each sampled housing unit constantly represents about 2,000 households in the metropolitan areas. Because the AHS data were collected from the major metropolitan areas in the whole country for analysis, crash data have to be nationwide as well. Although the FARS provides only fatal crashes and is possibly biased, it is the only nationwide crash database. Thus, it was decided that the FARS data should be used for the empirical analysis. The numbers of fatal pedestrian and bicycle crashes were extracted from the FARS database. The spatial distributions of fatal pedestrian and bicycle crashes are presented in Figures 3 and 4, respectively. The data from the gray areas in the maps were not available from the AHS data, and not considered in this study. The descriptive statistics of the prepared data are summarized in Table 1.

Results and Discussion

Tables 2 and 3 present the integrated and non-integrated bivariate models for pedestrian/bicycle trips and crashes, respectively. In terms of overall DIC, the integrated model considerably outperforms the non-integrated counterpart (186.492 versus 262.830) as Spiegelhalter stated that differences over 10 might definitely rule out the model with the larger DIC (*32*). In the probability prediction model, the difference of DIC between the integrated and non-integrated structures is less than five (60.001 vs. 56.664). If the difference is less than five, it would be difficult to say that they have a meaningful difference in the model performance (*32*). On the other hand, there is a huge difference in the crash prediction model component between the integrated and non-integrated approaches (126.491 vs. 206.186). It clearly shows that the integrated modeling approach has a significantly better performance compared with its non-integrated counterpart. It implies that the integrated approach is more efficient than both model components because the crash prediction models could utilize additional information from the probability models. The following result interpretation is based on the integrated bivariate model (Table 2).

Pedestrian Use Probability Model

In the pedestrian trip probability prediction model, overall four variables were found to be significant. It was found that a higher proportion of overall opinion of present household structure is poor and higher median household income would decrease the probability of choosing a walking mode. Meanwhile, the proportion of pedestrian accessibility to a grocery store has a positive

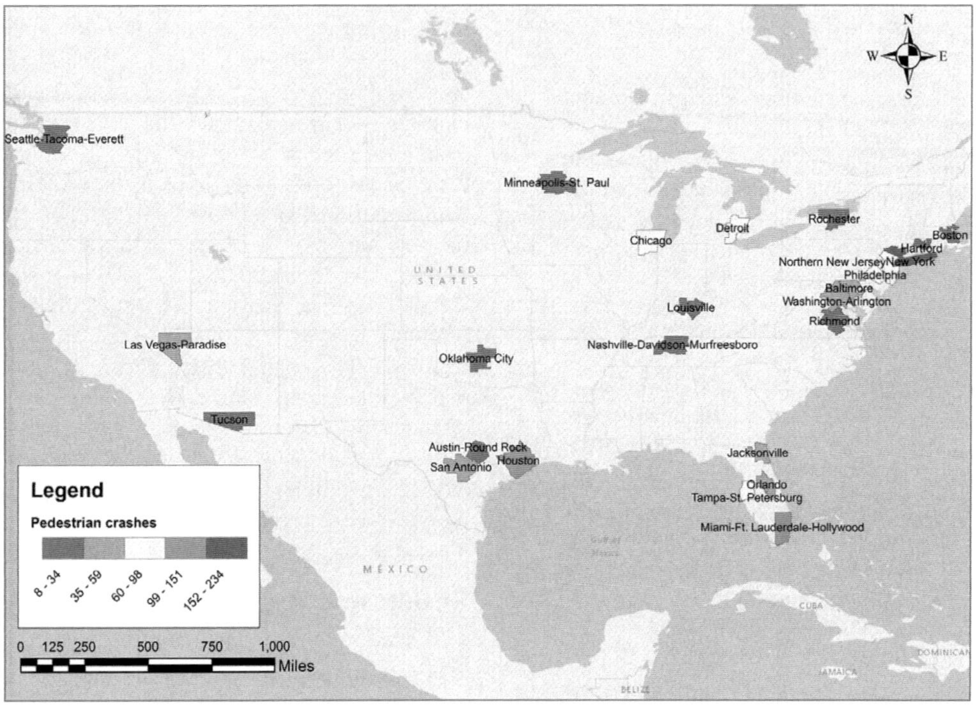

Figure 3. Spatial distribution of pedestrian fatal crashes by AHS area.

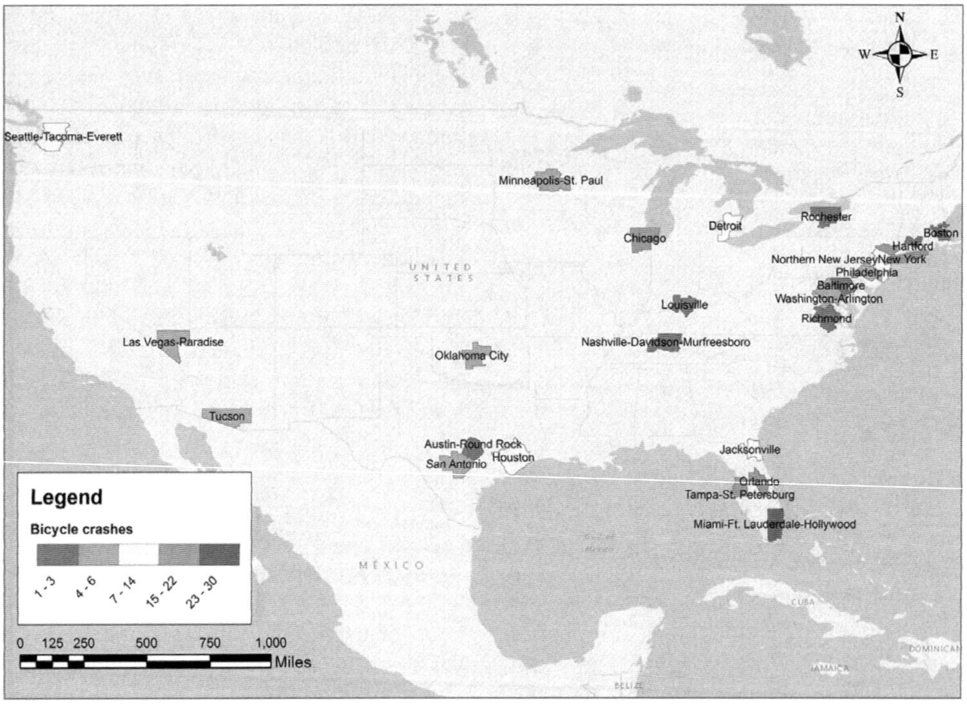

Figure 4. Spatial distribution of bicycle fatal crashes by AHS area.

Table 1. Descriptive Statistics of the Selected Variables

Category	Variable	Mean	SD	Min	Max
Trips	Probability of choosing a bicycle mode	0.035	0.011	0.017	0.057
	Probability of choosing a walking mode	0.268	0.077	0.152	0.491
Crashes	Fatal crashes involving a pedestrian	58.200	51.436	8	234
	Fatal crashes involving a bicycle	8.320	7.598	1	30
Household characteristics	Total occupied housing units	1285.5	954.9	373	4305
	Proportion of new construction past 4 years	0.018	0.011	0.006	0.042
	Proportion of manufactured/mobile homes	0.036	0.028	0.001	0.098
	Proportion of black householders	0.150	0.074	0.035	0.286
	Proportion of Hispanic householders	0.150	0.121	0.025	0.491
	Proportion of elderly (65 years and over)	0.221	0.031	0.140	0.284
	Proportion of below poverty level	0.152	0.029	0.095	0.190
	Proportion of vehicles available	0.907	0.072	0.587	0.955
	Proportion of one vehicle available	0.340	0.035	0.283	0.407
	Proportion of two vehicles available	0.369	0.043	0.190	0.414
	Proportion of three or more vehicles available	0.198	0.038	0.112	0.260
	Proportion of no vehicle available	0.093	0.072	0.045	0.413
	Proportion of overall opinion of present structure is poor	0.015	0.004	0.008	0.023
	Proportion of overall opinion of present structure is fair	0.249	0.020	0.209	0.289
	Proportion of overall opinion of present structure is good	0.736	0.021	0.693	0.779
	Median household income (1,000 USD)	53.749	11.193	41	85
Pedestrian accessibility	Proportion of grocery store	0.084	0.047	0.036	0.266
	Proportion of personal services	0.072	0.049	0.030	0.263
	Proportion of retail shopping	0.062	0.040	0.028	0.214
	Proportion of entertainment	0.066	0.037	0.029	0.197
	Proportion of health care services	0.043	0.033	0.013	0.165
	Proportion of personal banking	0.058	0.044	0.017	0.218
	Proportion of school or workplace	0.031	0.017	0.013	0.089
Bicycle accessibility	Proportion of grocery store	0.026	0.011	0.012	0.062
	Proportion of personal services	0.020	0.009	0.007	0.041
	Proportion of retail shopping	0.022	0.011	0.009	0.053
	Proportion of entertainment	0.018	0.009	0.007	0.036
	Proportion of health care services	0.021	0.012	0.006	0.055
	Proportion of personal banking	0.022	0.012	0.006	0.064
	Proportion of school or workplace	0.023	0.012	0.008	0.053
Public transportation accessibility	Proportion of grocery store	0.557	0.111	0.353	0.779
	Proportion of personal services	0.535	0.112	0.339	0.740
	Proportion of retail shopping	0.918	0.026	0.852	0.950
	Proportion of entertainment	0.061	0.018	0.038	0.118
	Proportion of health care	0.021	0.009	0.006	0.038
	Proportion of personal banking	0.520	0.113	0.327	0.730
Reason for not walking or cycling	Proportion of health reasons	0.080	0.020	0.042	0.115
	Proportion of lack of adequate sidewalks	0.039	0.023	0.005	0.095
	Proportion of crime and safety concerns	0.018	0.008	0.004	0.037
	Proportion of traffic issues	0.028	0.016	0.004	0.074
	Proportion of no bike lanes	0.023	0.017	0.005	0.082
	Proportion of no place close enough	0.037	0.020	0.008	0.082
	Proportion of no time	0.024	0.010	0.010	0.056
	Proportion of no bike	0.033	0.012	0.006	0.067
	Proportion of not enough light	0.011	0.009	0.002	0.045
Land-use (area within half block)	Proportion of open space, park, woods, farm, or ranch	0.389	0.057	0.294	0.469
	Proportion of 4-or-more-lane highway, railroad, or airport	0.137	0.029	0.082	0.199
	Proportion of commercial or institutional	0.304	0.084	0.180	0.535
	Proportion of industrial structures or factories	0.040	0.019	0.013	0.096
	Proportion of with parking lots	0.251	0.040	0.175	0.320
Structure or street characteristics (area within half block)	Proportion of no bars on windows	0.881	0.098	0.580	0.968
	Proportion of one vandalized or abandoned building	0.033	0.009	0.017	0.049
	Proportion of more than one vandalized or abandoned building	0.039	0.022	0.009	0.103
	Proportion of small amount of trash, litter, or junk in streets, or on any properties	0.061	0.018	0.038	0.118
	Proportion of large amount of trash, litter, or junk in streets, or on any properties	0.021	0.009	0.006	0.038
	Proportion of streets need minor repairs	0.306	0.064	0.192	0.418
	Proportion of streets need major repairs	0.059	0.028	0.027	0.133

(continued)

Table 1. Continued

Category	Variable	Mean	SD	Min	Max
Police protection	Proportion of satisfactory police protection	0.911	0.024	0.862	0.957
	Proportion of unsatisfactory police protection	0.070	0.021	0.032	0.121
Road facility	Proportion of sidewalks present: adequately lit	0.504	0.145	0.241	0.758
	Proportion of sidewalks present: not adequately lit	0.112	0.051	0.057	0.298
	Proportion of sidewalks not present	0.371	0.145	0.159	0.672
	Proportion of bike lanes present	0.162	0.083	0.053	0.321
	Proportion of bike lanes not present	0.827	0.086	0.671	0.948
Households using public transportation	Proportion of always	0.250	0.077	0.158	0.486
	Proportion of most of the time	0.098	0.020	0.064	0.138
	Proportion of sometimes	0.101	0.022	0.067	0.140
	Proportion of once in a while	0.143	0.034	0.061	0.200
	Proportion of never	0.404	0.060	0.220	0.525
Main mode to get to public transportation	Proportion of walk	0.672	0.114	0.458	0.882
	Proportion of local bus	0.128	0.052	0.038	0.219
	Proportion of drive or someone else drive	0.149	0.094	0.037	0.311
	Proportion of Metrorail	0.008	0.012	0.000	0.046
	Proportion of bike	0.008	0.009	0.000	0.038
	Proportion of commuter bus or shuttle van	0.009	0.006	0.001	0.024
	Proportion of commuter or intercity train	0.004	0.005	0.000	0.019

Note: SD = standard deviation.

Table 2. Integrated Bivariate Modeling Result for Pedestrian/Bicycle Use Probability and Crashes

				Bayesian Confidence Interval			
Category	Variable	Mean	SD	2.5%	5.0%	95.0%	97.5%
Walking Probability Prediction Model							
	Intercept	−2.710	0.473	−3.642	−3.450	−1.924	−1.785
Household characteristics	Proportion of overall opinion of present structure is poor	−36.870	1.004	−38.840	−38.520	−35.220	−34.860
	Median household income (1,000 USD)	−0.015[#]	0.009	−0.034	−0.031	−0.002	0.001
Pedestrian accessibility	Proportion of grocery store	3.301	0.941	1.455	1.776	4.848	5.181
Reason for not walking or cycling	Proportion of not enough light	−6.882	1.003	−8.849	−8.528	−5.243	−4.920
Bicycling Probability Prediction Model							
	Intercept	−5.134	0.354	−5.857	−5.734	−4.558	−4.466
Household characteristics	Proportion of black householders	1.589[#]	0.868	−0.118	0.155	3.035	3.308
	Proportion of no vehicle available	1.582	0.781	0.062	0.301	2.869	3.135
Bicycle accessibility	Proportion of grocery store	14.210	1.006	12.200	12.540	15.850	16.180
	Proportion of school or workplace	13.810	0.999	11.870	12.170	15.450	15.780
Land-use (area within half block)	Proportion of industrial structures or factories	1.978	0.991	0.049	0.324	3.597	3.900
Reason for not walking or cycling	Proportion of lack of adequate sidewalks	−5.333	0.987	−7.279	−6.963	−3.709	−3.410
	Proportion of no time	−10.130	1.004	−12.080	−11.750	−8.469	−8.146
Pedestrian Crash Prediction Model							
	Intercept	0.738	0.279	0.189	0.272	1.195	1.284
Exposure	Log of the "expected" pedestrian trips	0.832	0.091	0.662	0.689	0.981	1.017
Household characteristics	Proportion of black householders	2.971	0.307	2.352	2.463	3.477	3.573
	Proportion of Hispanic householders	1.348	0.284	0.791	0.877	1.813	1.904
	Proportion of below poverty level	6.261	0.319	5.626	5.733	6.786	6.884
Land-use (area within half block)	Proportion of open space, park, woods, farm, or ranch	−2.962	0.305	−3.555	−3.457	−2.457	−2.360
Bicycle Crash Prediction Model							
	Intercept	0.396[##]	0.481	−0.561	−0.406	1.190	1.329
Exposure	Log of the "expected" bicycle trips	0.858	0.120	0.626	0.665	1.056	1.101
Household characteristics	Proportion of manufactured/mobile homes	17.230	0.972	15.320	15.630	18.840	19.120

(continued)

Table 2. Continued

Category	Variable	Mean	SD	Bayesian Confidence Interval 2.5%	5.0%	95.0%	97.5%
Land-use (area within half block)	Proportion of open space, park, woods, farm, or ranch	−6.322	0.806	−7.959	−7.663	−5.011	−4.769
	Proportion of 4-or-more-lane highway, railroad, or airport	7.883	0.965	5.972	6.271	9.465	9.738
DIC	Pedestrian/bicycle use probability prediction model	60.001					
	Pedestrian/bicycle crash prediction model	126.491					
	Total	186.492					

Note: SD = standard deviation, ## = not significant at 90% BCI, # = only significant at 90% BCI, and all other variables are significant at 95% BCI.

Table 3. Non-Integrated Bivariate Modeling Result for Pedestrian/Bicycle Use Probability and Crashes

Category	Variable	Mean	SD	Bayesian Confidence Interval 2.5%	5.0%	95.0%	97.5%
Walking Probability Prediction Model							
	Intercept	−0.840##	0.975	−2.733	−2.436	0.787	1.094
Household characteristics	Proportion of overall opinion of present structure is poor	−36.910	1.004	−38.890	−38.560	−35.230	−34.930
	Median household income (1,000 USD)	−0.080	0.022	−0.126	−0.118	−0.044	−0.038
Pedestrian accessibility	Proportion of grocery store	4.351	1.017	2.349	2.687	6.047	6.361
Reason for not walking or cycling	Proportion of not enough light	−6.751	1.019	−8.760	−8.430	−5.066	−4.747
Bicycling Probability Prediction Model							
	Intercept	−4.926	0.364	−5.687	−5.552	−4.351	−4.245
Household characteristics	Proportion of black householders	0.949##	0.979	−0.974	−0.661	2.569	2.882
	Proportion of no vehicle available	1.501##	0.979	−0.444	−0.114	3.105	3.380
Bicycle accessibility	Proportion of grocery store	14.190	0.994	12.230	12.550	15.830	16.120
	Proportion of school or workplace	13.780	1.006	11.800	12.120	15.440	15.750
Land-use (area within half block)	Proportion of industrial structures or factories	2.191	0.997	0.233	0.524	3.821	4.160
Reason for not walking or cycling	Proportion of lack of adequate sidewalks	−4.999	0.991	−6.914	−6.605	−3.349	−3.053
	Proportion of no time	−10.100	0.993	−12.080	−11.740	−8.501	−8.177
Pedestrian Crash Prediction Model							
	Intercept	0.083##	0.267	−0.440	−0.365	0.526	0.601
Exposure	Log of the number of "expected" pedestrian trips	0.602	0.050	0.506	0.522	0.685	0.701
Household characteristics	Proportion of black householders	2.959	0.299	2.364	2.465	3.445	3.541
	Proportion of Hispanic householders	1.314	0.266	0.788	0.875	1.752	1.841
	Proportion of below poverty level	6.221	0.313	5.611	5.712	6.736	6.823
Land-use (area within half block)	Proportion of open space, park, woods, farm, or ranch	−3.279	0.296	−3.866	−3.767	−2.795	−2.694
Bicycle Crash Prediction Model							
	Intercept	−0.915##	0.655	−2.150	−1.975	0.209	0.431
Exposure	Log of the number of "expected" bicycle trips	0.725	0.119	0.472	0.516	0.909	0.943
Household characteristics	Proportion of manufactured/mobile homes	16.810	0.976	14.910	15.190	18.420	18.710
Land-use (area within half block)	Proportion of open space, park, woods, farm, or ranch	−6.842	0.836	−8.506	−8.229	−5.495	−5.214
	Proportion of 4-or-more-lane highway, railroad, or airport	7.860	0.973	5.952	6.270	9.456	9.773
DIC	Pedestrian/bicycle use probability prediction model	56.664					
	Pedestrian/bicycle crash prediction model	206.186					
	Total	262.830					

Note: SD = standard deviation, ## = not significant at 90% BCI, # = only significant at 90% BCI, and all other variables are significant at 95% BCI.

relationship with the pedestrian trip probability. Lastly, the proportion of households whose reason for not walking or cycling is insufficient street lighting also decreases the pedestrian trip probability.

Bicycle Use Probability Model

The bicycle probability model revealed seven significant variables. Both higher proportions of black householders and no vehicle available increase the bicycle use probability. The proportion of bicycle accessibilities to grocery store and school/workplace is positively related. If the industrial structures or factories are within half a block, households are more likely to use a bicycle. On the other hand, if the reasons for not walking or cycling are lack of adequate sidewalks or no time, households in the area have a lower propensity to use a bicycle.

Pedestrian Crash Prediction Model

In the integrated model, the natural logarithm of expected pedestrian trips was used as an exposure. As expected it has a positive effect on the number of fatal pedestrian crashes. It was shown that three household variables have a positive relationship: the proportions of black householders, Hispanic householders, and below poverty level. It was shown that the proportion of open space, park, woods, farm, or ranch is negatively related.

Bicycle Crash Prediction Model

Similarly in the pedestrian crash prediction model part, the natural logarithm of expected bicycle trips was included as an exposure variable and it is positively related with the number of fatal bicycle crashes. A household in the area with the larger number of manufactured/mobile homes is more likely to be involved in bicycle crashes. Two land-use variables were found significant. If an area has a higher proportion of open space, park, woods, farm, or ranch, households in the area are less likely to be involved in fatal bicycle crashes. In contrast, households in the area with a higher proportion of 4-or-more-lane highway, railroad, or airport have a higher tendency to be involved in bicycle crashes.

The contributing factors in the walking/bicycling probability models would help practitioners to establish an effective strategy to encourage the non-motorized modes. For example, to promote walking and cycling, it is suggested to provide a better accessibility to certain facilities (e.g., grocery store), and supply sufficient lighting and sidewalks. It is shown that the areas with a lower median household income and more households with no vehicle have a higher probability of using non-motorized modes. It implies that there is a perception that walking and bicycling are only for economically deprived people. Transportation planners should focus on the areas that are not deprived to provide education and campaign to promote walking and bicycling. On the other hand, the contributing factors in the pedestrian and bicycle crash models would be useful for practitioners to develop a successful plan to reduce non-motorized crashes. For instance, it is recommended to focus on the areas with specific race/ethnic groups, economically deprived households (e.g., below poverty level, manufactured/mobile homes), and higher transportation activities (e.g., 4-or-more lane highways, railroad, or airport). It is recommended to provide safety education, campaigns, engineering countermeasures, frequent enforcement/patrol, and so on to decrease the number of non-motorized mode involved crashes. Readers should note that the pedestrian/bicycle trips were used as an exposure (or control) variable for crashes.

Summary and Conclusions

In recent years, considerable efforts have been made to incorporate safety into LTTPs, which is termed transportation safety planning (TSP). Federal laws require that state DOTs' and MPOs' planning processes should be consistent with Strategic Highway Safety Plans (*1*), such as the MAP-21 Act and FAST Act. In this context of TSP, many efforts have been put into integrating planning and safety. Although some macro-level safety studies have attempted to adopt planning data (e.g., trip generation) for estimating traffic crash frequency, no studies have developed trip and crash models simultaneously. In this study, the authors suggested an integrated approach for modeling pedestrian and bicycle trips and crashes simultaneously in one structure. The AHS data were collected from the U.S. Census Bureau and used to validate the proposed method. The AHS is a longitudinal housing unit survey that asks questions about the quality of housing including but not limited to demographic, socioeconomic, walking or cycling environments, public transportation systems, and accessibility to various facilities, in major metropolitan areas.

In the integrated modeling structure, the first component is a logistic regression model, which estimates the probability of pedestrians or bicyclists' trips among total trips in each metropolitan area. The second part is a Poisson lognormal regression model, which estimates pedestrian or bicycle-involved crash counts. In the crash models, the expected number of trips from the first component was fed into crash prediction models. In addition, a multivariate modeling structure was implemented in the first and second model components. Multiple significant variables were found to be significant for the probability of choosing walking or cycling, and the number of crashes involving pedestrians or bicycles. For the probability of choosing non-motorized modes, the perceived household structure, income, accessibility to grocery store or school/workplace, and walking/cycling facilities play important roles. For the crash prediction models, race (e.g., black, Hispanic), the specific type of house structure (e.g., manufactured/mobile homes) poverty, and land-use have significant impacts. In the crash models, estimated pedestrian and bicycle trips were used as exposure (control) variables, which can explain most of the variance in the crash counts. It does not mean that the planners should reduce the number of non-motorized mode crashes by discouraging pedestrian or bicycle trips. All the other contributing factors that were identified from the study can be used for policy makers to focus on the specific areas with those characteristics (e.g., higher proportions of households below poverty level, black/Hispanic population, etc.), and put more efforts and resources into minimizing traffic crashes by providing education, safety campaigns, engineering countermeasures, frequent enforcement/patrol, and so on.

In addition, it was found that the integrated approach significantly outperforms its non-integrated counterpart. It may imply that both components could utilize supplementary information from each other in the integrated modeling structure. To the authors' best knowledge, this study is the first one which has attempted to combine trip generation and crash models in one modeling structure. It is expected that the integrated modeling approach for trips and crashes in this study will provide great insights into the future directions of TSP.

Author Contributions

The authors confirm contribution to the paper as follows: study conception and design: Jaeyoung Lee, Mohamed Abdel-Aty, Helai Huang; data collection: Jaeyoung Lee; analysis and interpretation of results: Jaeyoung Lee, Qing Cai, Ling Wang; draft manuscript preparation: Jaeyoung Lee. All authors reviewed the results and approved the final version of the manuscript.

References

1. Federal Highway Administration. *Transportation Safety Planning*. U.S. Department of Transportation, Washington, D.C. https://safety.fhwa.dot.gov/tsp. Accessed on April 17, 2018, 2018.
2. Washington, S., I. Van Schalkwyk, S. Mitra, M. Mayer, E. Dumbaugh, and M. Zoll. *Incorporating Safety into Long-Range Transportation Planning*. NCHRP Report 546. Transportation Research Board, Washington, D.C., 2006.
3. Lovegrove, G. R., and T. Sayed. Macro-Level Collision Prediction Models for Evaluating Neighbourhood Traffic Safety. *Canadian Journal of Civil Engineering*, Vol. 33, No. 5, 2006, pp. 609–621.
4. Lovegrove, G., and T. Sayed. Using Macrolevel Collision Prediction Models in Road Safety Planning Applications. *Transportation Research Record: Journal of the Transportation Research Board*, 2006. 1950: 73–82.
5. Naderan, A., and J. Shahi. Aggregate Crash Prediction Models: Introducing Crash Generation Concept. *Accident Analysis & Prevention*, Vol. 42, No. 1, 2010, pp. 339–346.
6. Abdel-Aty, M., C. Siddiqui, H. Huang, and X. Wang. Integrating Trip and Roadway Characteristics to Manage Safety in Traffic Analysis Zones. *Transportation Research Record: Journal of the Transportation Research Board*, 2011. 2213: 20–28.
7. Pirdavani, A., T. Brijs, T. Bellemans, B. Kochan, and G. Wets. Application of Different Exposure Measures in Development of Planning-Level Zonal Crash Prediction Models. *Transportation Research Record: Journal of the Transportation Research Board*, 2012. 2280: 145–153.
8. Lee, J., M. Abdel-Aty, and X. Jiang. Development of Zone System for Macro-Level Traffic Safety Analysis. *Journal of Transport Geography*, Vol. 38, 2014, pp. 13–21.
9. Abdel-Aty, M., J. Lee, C. Siddiqui, and K. Choi. Geographical Unit Based Analysis in the Context of Transportation Safety Planning. *Transportation Research Part A: Policy and Practice*, Vol. 49, 2013, pp. 62–75.
10. Xu, P., H. Huang, N. Dong, and M. Abdel-Aty. Sensitivity Analysis in the Context of Regional Safety Modeling: Identifying and Assessing the Modifiable Areal Unit Problem. *Accident Analysis & Prevention*, Vol. 70, 2014, pp. 110–120.
11. Cai, Q., M. Abdel-Aty, J. Lee, and N. Eluru. Comparative Analysis of Zonal Systems for Macro-Level Crash Modeling. *Journal of Safety Research*, Vol. 61, 2017, pp. 157–166.
12. Lee, J., M. Abdel-Aty, and K. Choi. Analysis of Residence Characteristics of At-Fault Drivers in Traffic Crashes. *Safety Science*, Vol. 68, 2014, pp. 6–13.
13. Dong, N., H. Huang, J. Lee, M. Gao, and M. Abdel-Aty. Macroscopic Hotspots Identification: A Bayesian Spatio-Temporal Interaction Approach. *Accident Analysis & Prevention*, Vol. 92, 2016, pp. 256–264.
14. Siddiqui, C., and K. Watkins. Evaluating Long-Range Regional Safety with Scenario Planning Analysis. *Transportation Research Record: Journal of the Transportation Research Board*, 2016. 2563: 18–28.
15. Wier, M., J. Weintraub, E. H. Humphreys, E. Seto, and R. Bhatia. An Area-Level Model of Vehicle-Pedestrian Injury Collisions with Implications for Land Use and Transportation Planning. *Accident Analysis & Prevention*, Vol. 41, No. 1, 2009, pp. 137–145.
16. Siddiqui, C., M. Abdel-Aty, and K. Choi. Macroscopic Spatial Analysis of Pedestrian and Bicycle Crashes. *Accident Analysis & Prevention*, Vol. 45, 2012, pp. 382–391.
17. Lee, J., M. Abdel-Aty, K. Choi, and H. Huang. Multi-Level Hot Zone Identification for Pedestrian Safety. *Accident Analysis & Prevention*, Vol. 76, 2015, pp. 64–73.
18. Cai, Q., J. Lee, N. Eluru, and M. Abdel-Aty. Macro-Level Pedestrian and Bicycle Crash Analysis: Incorporating Spatial Spillover Effects in Dual State Count Models. *Accident Analysis & Prevention*, Vol. 93, 2016, pp. 14–22.

19. Wang, X., J. Yang, C. Lee, Z. Ji, and S. You. Macro-Level Safety Analysis of Pedestrian Crashes in Shanghai, China. *Accident Analysis & Prevention*, Vol. 96, 2016, pp. 12–21.
20. Amoh-Gyimah, R., M. Saberi, and M. Sarvi. Macroscopic Modeling of Pedestrian and Bicycle Crashes: A Cross-Comparison of Estimation Methods. *Accident Analysis & Prevention*, Vol. 93, 2016, pp. 147–159.
21. Xie, K., K. Ozbay, A. Kurkcu, and H. Yang. Analysis of Traffic Crashes Involving Pedestrians Using Big Data: Investigation of Contributing Factors and Identification of Hotspots. *Risk Analysis*, Vol. 37, No. 8, 2017, pp. 1459–1476
22. Shah, M. I., M. Abdel-Aty, and J. Lee. Evaluation of Surrogate Measures for Pedestrian Exposure to Traffic Crashes at Intersections. Presented at 96th Annual Meeting of the Transportation Research Board, Washington, D.C., 2017.
23. Lee, J., S. Yasmin, N. Eluru, M. Abdel-Aty, and Q. Cai. Analysis of Crash Proportion by Vehicle Type at Traffic Analysis Zone Level: A Mixed Fractional Split Multinomial Logit Modeling Approach with Spatial Effects. *Accident Analysis & Prevention*, Vol. 111, 2018, pp. 12–22.
24. Lee, J., and M. Abdel-Aty. Macro-Level Analysis of Bicycle Safety: Focusing on the Characteristics of Both Crash Location and Residence. *International Journal of Sustainable Transportation*, 2017, Vol. 0, pp. 1–8.
25. Wang, L., M. Abdel-Aty, and J. Lee. Safety Analytics for Integrating Crash Frequency and Real-Time Risk Modeling for Expressways. *Accident Analysis & Prevention*, Vol. 104, 2017, pp. 58–64.
26. Lee, J., M. Abdel-Aty, and X. Jiang. Multivariate Crash Modeling for Motor Vehicle and Non-Motorized Modes at the Macroscopic Level. *Accident Analysis & Prevention*, Vol. 78, 2015, pp. 146–154.
27. Ye, X., R. M. Pendyala, S. P. Washington, K. Konduri, and J. Oh. A Simultaneous Equations Model of Crash Frequency by Collision Type for Rural Intersections. *Safety Science*, Vol. 47, No. 3, 2009, pp. 443–452.
28. Yu, R., and M. Abdel-Aty. Investigating Different Approaches to Develop Informative Priors in Hierarchical Bayesian Safety Performance Functions. *Accident Analysis & Prevention*, Vol. 56, 2013, pp. 51–58.
29. Congdon, P. *Applied Bayesian Modelling*. John Wiley & Sons, 2014.
30. Spiegelhalter, D. J., N. G. Best, B. P. Carlin, and A. Van Der Linde. Bayesian Measures of Model Complexity and Fit. *Journal of the Royal Statistical Society: Series B (Statistical Methodology)*, Vol. 64, No. 4, 2002, pp. 583–639.
31. U.S. Census Bureau. *American Housing Survey*. https://www.census.gov/programs-surveys/ahs/about.html. Accessed June 29, 2017.
32. Spiegelhalter, D., A. Thomas, N. Best, and D. Lunn. *WinBUGS User Manual*. MRC Biostatistics Unit, Cambridge. https://www.mrc-bsu.cam.ac.uk/wp-content/uploads/manual14.pdf. Accessed May 29, 2017. U.S. Department of Commerce

The Standing Committee on Transportation Safety Management (ANB10) peer-reviewed this paper (18-01759).

Article

TRR
JOURNAL OF THE TRANSPORTATION RESEARCH BOARD

Transportation Research Record
2018, Vol. 2672(32) 61–71

DOI: 10.1177/0361198118772724
journals.sagepub.com/home/trr
SAGE

Analysis of Fatal Traffic Crash-Reporting and Reporting-Arrival Time Intervals of Emergency Medical Services

Jaeyoung Lee[1], Mohamed Abdel-Aty[2], Qing Cai[2], and Ling Wang[2,3]

Abstract

Emergency Medical Services (EMS) play a vital role in the postcrash effort to reduce fatalities by providing first aid and transportation to medical facilities. This study aims to analyze the time required for crash reporting and EMS arrival in fatal traffic crashes and to identify relevant crash, roadway, environmental and zonal socio-economic factors. The time required for EMS reporting and arrivals were calculated by location type (urban or rural) and roadway functional classification using Florida data. Subsequently, a variety of duration models were estimated to reveal the contributing factors for crash-reporting and reporting-arrival intervals. Although about 90% of fatal crashes are reported to EMS within ten minutes in both urban and rural settings, EMS average reporting time in rural areas (4.5 min) is greater than that in urban areas (3 min). Moreover, freeways require longer time for EMS arrival (8.3 min) compared with conventional roadways (6.8 min). It was shown that the log-logistic and gamma models performed the best for crash-reporting and reporting-arrival intervals, respectively. The modeling results revealed that both EMS reporting and arrival times are related to the crash, roadway, environmental, and socio-economic factors. The key findings indicate that EMS reporting and arrival times differ significantly according to the urban or rural designation and road functional classification, and that they have statistically significant relationship with various factors. It is expected that the findings from this study can be used to develop effective and practical strategic plans to minimize EMS reporting and arrival times and, therefore, decrease the likelihood of fatalities.

According to the United States National Highway Transportation Safety Administration (NHTSA), close to 30,000 people are killed from traffic crashes every year (*1*). In 2014, slightly more than 10 out of every 100,000 of the population were killed from traffic crashes, which indicate that traffic crashes are one of the major public health concerns. Over the last several decades, many efforts have been made to reduce traffic fatalities. One of the efforts is to quickly and efficiently dispatch Emergency Medical Services (EMS) to the crash scene. EMS has played a vital role in reducing fatalities by providing first aid and transportation to medical facilities as one of the postcrash countermeasures.

According to Clark et al. (*2*), most deaths caused by traffic crashes happen before the patient has arrived at a hospital, indicating that efficient EMS can reduce the impact of traffic crashes. Especially, EMS has been proven to be effective in minimizing the fatality rates (*3, 4*). In general, the time interval between crash occurrence and reporting (T1) and the time interval between reporting and EMS arrival on the scene (T2) are the two major factors for EMS (*2, 3, 5, 6*). Several studies investigated the relationship between the outcome of traffic crashes and the time intervals (i.e., T1 and T2). Castrén et al. (*7*) found that fatalities would be reduced by 6% as T1 is reduced by 1 minute. Also, Sánchez-Mangas et al. (*4*) uncovered that a 10 minute reduction in T2 can decrease the probability of traffic fatalities by one-third.

Despite of the importance of rapid EMS response, only a few studies have explored the contributing factors for the time required for reporting and arrival. Brodsky (*3*) explored the factors for delays in reporting time of EMS in traffic crashes. Three explanatory variables,

[1]Center for Advanced Transportation Systems Simulation (CATSS), Department of Civil, Environmental, and Construction Engineering, University of Central Florida, Orlando, FL
[2]Department of Civil, Environmental, and Construction Engineering, University of Central Florida, Orlando, FL
[3]College of Transportation Engineering, Tongji University, Shanghai, China

Corresponding Author:
Address correspondence to Jaeyoung Lee: Jaeyoung@knights.ucf.edu

population density, traffic density, and hour of day, were tested for association with an extremely delayed reporting time using logistic regression. Other factors, such as crash, road type, and severity of injury, showed little association with delays. The author also found that notification delays are likely to be accompanied by longer ambulance response time as ambulance stations cover broad areas in remote counties. Yasunaga et al. (*8*) analyzed the time between emergency call and ambulance arrival using population density for cardiac arrest cases. The authors found that the mean call-response intervals in low-density and high-density were 9.3 and 6.2 min, respectively. The distribution of EMS centers was almost proportional to the size of the population. EMS resource allocation according to population size may cause disparities in response times. Increasing the number of ambulances could decrease call-response intervals across large, sparsely populated areas. Evanco (*9*) and Gonzalez et al. (*10*) identified that longer T1 and T2 can be observed in the crashes that occurred in rural areas. Brodsky (*11*) found that crashes occurring during nighttime and on interstate highways need more time to be reported. In the study of Meng and Weng (*12*), it was shown that crashes happening on holidays and those with poor light condition may have longer T1. Meng and Weng (*12*) and Yasmin et al. (*13*) commonly found that the time of crash occurrence, road type, holiday, and weather were found to be significant factors for T2.

Nevertheless, these studies attempted a limited number of variables to explain EMS reporting and arrival time, such as crash characteristics, population density, and so forth. Therefore, this study aims to analyze the time required for EMS reporting and dispatch in fatal traffic crashes using various explanatory factors including crash, roadway, environmental, and zonal socioeconomic factors using duration models. Also, several suggestions to shorten EMS reporting and arrival time intervals are provided based on the modeling results.

Methods

Study Design

The research team defined the time interval between crash occurrence and notification to EMS as T1 (i.e., crash-reporting interval); and the time interval between the notification to EMS and EMS arrival on the scene as T2 (i.e., reporting-arrival interval) as shown in Figure 1. The overall average crash-reporting and reporting-arrival intervals were calculated and then the average crash-reporting and reporting-arrival intervals by geographic locations (urban or rural) and roadway functional classifications were computed. Furthermore, two duration

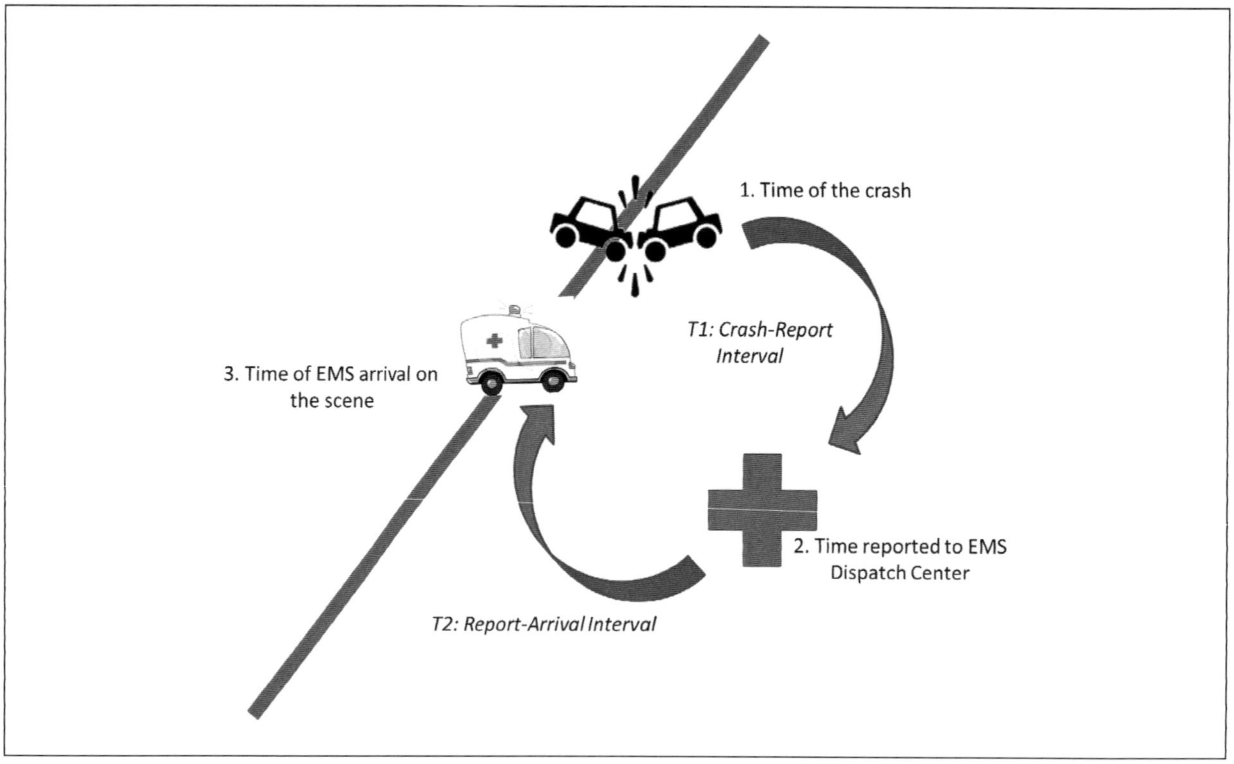

Figure 1. Model estimation process of the integrated approach.

models were developed to identify significant factors for crash-reporting and reporting-arrival intervals. Hazard-based duration models have been extensively applied to the public health, medical/biometrics, industrial, and transportation engineering fields. The variable of interest in duration models is survival time elapsed from the beginning of an event until its ending. In this study, crash-reporting and reporting-arrival intervals were used as two targets of the duration models. The distributional form of the error term determines the duration regression model. Five distributions were attempted for crash-reporting and reporting-arrival intervals: exponential, gamma, log-logistic, log-normal, and Weibull. The models with the best fit were determined as the best models and utilized to reveal significant factors for the reporting and arrival times, respectively.

Study Population and Data

To achieve the objectives of this study, various data were collected from multiple sources. Six years of fatal crash and associated data (2005–2010) in Florida were collected from the Fatal Analysis Reporting System (FARS). These include information of crash location, time, occupant(s), roadway functional classification, light conditions, lane counts, weather, and so forth. Traffic Analysis Zones (TAZ) data were collected from the Florida Department of Transportation such as employment, school enrollment, car-ownership, traffic volume, roadway density, and so forth. Furthermore, population and commuter characteristics data were obtained from the U.S. Census Bureau. Lastly, land-use data (e.g., agriculture, industry, mining, public, residential land-use) were acquired from the Florida Department of Revenue. The collected data were processed for exploratory analysis and developing duration models. The prepared data are summarized in Table 1.

Duration Modeling

The author adopted a duration model to estimate the two target variables: T1 and T2. Hazard-based duration models have been used in the fields of biometrics, economics, marketing, and industrial engineering as a means of determining causality in duration data, and also applied in the transportation field. In the transportation field, duration models have been used in crash analysis (*14, 15*), travel behavior (*16–19*), car ownership (*20–22*), travel delay (*23–25*), and highway visibility (*26*). Both Nam and Mannering (*14*) and Chung (*15*) explored the incident durations using hazard-based duration models. In this study, the authors adopted the duration model for investigating the crash-reporting and reporting- arrival time intervals. The variable of interest in a duration model is the survival time that elapsed from the beginning of an event until its end. In this study, duration time can be regarded as the duration that starts when a crash occurs and ends when it is reported (T1). Also, it can be considered as the waiting duration that starts when the crash is reported and ends when EMS vehicle arrives at the crash scene (T2). Therefore, a duration model is appropriately used in this study. To the authors' best knowledge, hazard-based duration models have not been used for EMS reporting or dispatch time analysis. For more detailed information for failure rate or hazard function, please refer to Bhat (*16*). The most commonly used approach to model duration data is the proportional hazard model. The proportional hazard models for duration data usually assume that the explanatory variables take a constant proportional effect on an unspecified baseline hazard function. Although this assumption may relieve the estimation efforts of the model, it may not be applicable when the constant proportional assumption is violated.

An alternative approach is the parametric accelerated hazard model, which is mainly used in reliability theory and industrial experiments. Contrary to the proportional hazard model, the accelerated hazard model is not necessarily limited to constant proportionality or additivity. Also, in the accelerated parametric hazard model, its hazard function can be flexibly chosen. On the other hand, non-parametric models can handle any shape of underlying hazard distribution including multiple peaks, which cannot be handled by the parametric model (*27*). Nevertheless, no such multiple peaks are assumed in the data of this study. For these reasons, a parametric accelerated hazard approach was adopted. In the parametric accelerated hazard model, the natural logarithm of the duration time, log t, is expressed as a linear function of the covariates, yielding the linear model as

$$\ln(t) = \mu + \varepsilon \tag{1}$$

where $\mu = x'\beta$, x is a vector of covariates, β is a vector of regression coefficients, and ε is the error with density $f(\varepsilon)$. It is noted that $(t_j + 1)$ was used for T1 instead of (t_j) in the equation as there are excessive zero values in T1, which means the crashes are reported within one minute. The variables that were statistically significant at 95% confidence level ($\alpha = 0.05$) were kept in the final models.

The distributional form of the error term determines the regression model. If we let $f(\varepsilon)$ be the logistic density, the log-logistic regression is obtained. If we let $f(\varepsilon)$ be the standard normal density, the log-normal regression is obtained. Setting $f(\varepsilon)$ equal to the extreme-value density yields the exponential and the Weibull regression models. Several parametric distributions can be assumed including exponential, gamma, log-logistic, log-normal, and

Table 1. Descriptive Statistics of the Prepared Data

Category	Variables	Mean	SD	Min.	Max.	Source
Target variables	T1: Crash-Reporting Interval (min)	3.642	8.252	0	120	NHTSA FARS
	T2: Reporting-Arrival Interval (min)	7.055	4.919	1	104	
Person	Number of killed persons in the crash	1.103	0.378	1	7	
	Number of total persons in the crash	2.636	1.888	1	57	
	Proportion of killed persons in the crash	0.560	0.288	0.018	1	
	Whether all crash-involved persons were killed[a]	0.265	0.441	0	1	
Traffic	Log (vehicle-miles-traveled per square mile)	9.235	2.065	0	15.373	FDOT
	Proportion of heavy vehicle traffic	0.074	0.053	0	0.428	
Functional classification	Freeway/expressway[a]	0.167	0.373	0	1	NHTSA FARS
	Arterial road[a]	0.465	0.499	0	1	
	Collector[a]	0.004	0.066	0	1	
Roadway	Lane counts	2.400	0.849	0	9	
	Curve[a]	0.186	0.389	0	1	
	Log (road length per square mile)	1.772	0.915	0	7.823	FDOT
Location	Intersection[a]	0.316	0.465	0	1	NHTSA FARS
	Interchange[a]	0.014	0.119	0	1	
	Ramp[a]	0.011	0.105	0	1	
	Urban[a]	0.591	0.492	0	1	
Driver action	Hit-and-run involved[a]	0.060	0.237	0	1	
Environment	Rain[a]	0.062	0.242	0	1	
	Fog[a]	0.011	0.106	0	1	
	Wet pavement[a]	0.112	0.315	0	1	
Lighting condition	Dark without lighting[a]	0.286	0.452	0	1	
	Dark with lighting[a]	0.277	0.448	0	1	
	Dawn[a]	0.019	0.138	0	1	
	Dusk[a]	0.024	0.152	0	1	
Land-use	Proportion of agriculture areas	0.109	0.238	0	0.994	FDOT
	Proportion of industrial areas	0.020	0.069	0	0.963	
	Proportion of institutional areas	0.020	0.063	0	0.986	
	Proportion of mining areas	0.001	0.016	0	0.537	
	Proportion of public areas	0.056	0.125	0	1	
	Proportion of recreational areas	0.029	0.094	0	0.929	
	Proportion of residential areas	0.178	0.229	0	1	
	Proportion of retail and office areas	0.048	0.100	0	1	
Car-ownership	Proportion of families with no vehicle	0.072	0.075	0	0.73	
	Proportion of families with 2 or more vehicles	0.540	0.163	0	1	
Socio-demographic	Log (population density)	6.473	2.027	0	10.837	
	Log (employment density)	5.572	2.232	0	12.074	
	Log (school enrollment density)	3.260	2.969	0	11.856	
	Proportion of industrial employments	0.207	0.237	0	1	
	Proportion of commercial employments	0.281	0.207	0	1	
	Proportion of service employments	0.505	0.234	0	1	
Commuter	Proportion of commuters using car	0.896	0.068	0	1	
	Proportion of commuters using public transit	0.020	0.037	0	0.549	
	Proportion of commuters using bicycle	0.006	0.013	0	0.234	
	Proportion of commuters who walk	0.016	0.025	0	0.437	
	Proportion of people working at home	0.045	0.036	0	0.807	

Note: SD = standard deviation.
[a]Variable is a binary dummy (yes = 1, no = 0).

Weibull. The mathematical expressions of the five error distributions are summarized in Table 2.

These five types of models were developed, of which the best models were determined based on mean absolute deviation (MAD), which is the mean of absolute differences between predicted and observed values. The equation to calculate the MAD is

$$MAD = \frac{\sum_{i=1}^{N} |obs_i - pred_i|}{N} \tag{2}$$

where obs_i is the observed value at i, $pred_i$ is the predicted value at i from the estimated model, N is the number of observations.

Table 2. Mathematical Expressions of the Error Distribution

Model	Functional form: $f(\varepsilon)$
Exponential	$f(\varepsilon) = \exp(w - \mu)\exp(-\exp(w - \mu))$
Gamma	$f(w) = \sigma^{-1} f'(u)$
	$f'(u) = \frac{\lvert\delta\rvert}{\Gamma(\delta^{-2})}\left(\delta^{-2}\exp(\delta u)\right)^{\delta^{-2}}$
	$\exp(-\exp(\delta u)\delta^{-2})$
Weibull	$f(\varepsilon) = \frac{1}{\sigma}\exp\left(\frac{w-\mu}{\sigma}\right)\exp\left(-\exp\left(\frac{w-\mu}{\sigma}\right)\right)$
Log-logistic	$f(\varepsilon) = \frac{\exp\left(\frac{w-\mu}{\sigma}\right)}{\sigma\left(1 + \exp\left(\frac{w-\mu}{\sigma}\right)\right)^2}$
Log-normal	$f(\varepsilon) = \frac{1}{\sqrt{2\pi}\sigma}\exp\left(-\frac{1}{2}\left(\frac{w-\mu}{\sigma}\right)^2\right)$

Results

Explanatory Analysis of Crash-Reporting and Reporting-Arrival Intervals

Overall 11,071 fatal crash cases were used in this study. The crash-reporting (T1) and reporting-arrival (T2) intervals were explored and the results are shown in Tables 3 and 4, respectively. Figure 2 shows the distributions of T1 and T2. It was revealed that the overall average T1 and T2 are 3.64 and 7.06 min, respectively. Approximately 94% of fatal crashes were reported within 10 min and EMS arrive on the crash scene within 10 min in 79% of the cases. Only in 21% of fatal crashes,

Table 3. Descriptive Statistics and Distribution of Crash-Reporting Interval (T1)

Descriptive Statistics

Location type	N	Mean	SD	Q1	Q2	Q3	Q4
Overall	11071	3.642	8.252	1	2	4	120
Urban	6546	3.027	7.555	0	1	3	120
Rural	4525	4.533	9.094	1	2	5	120
Freeway or expressway	1846	3.819	6.974	1	2	4	115
Conventional road	9225	3.607	8.484	1	2	4	120

Distribution

Required time	Overall	Urban	Rural	Freeway or expressway	Conventional road
5 min or less	79.8%	84.3%	73.2%	75.7%	80.6%
5–10 min	14.1%	11.4%	18.1%	17.3%	13.5%
11–15 min	2.79%	1.85%	4.15%	3.47%	2.66%
15–20 min	1.06%	0.73%	1.52%	1.57%	0.95%
21–25 min	0.43%	0.31%	0.62%	0.76%	0.37%
26–30 min	0.32%	0.24%	0.42%	0.27%	0.33%
More than 30 min	1.50%	1.17%	1.99%	0.93%	1.59%

Note: Q1 = First Quartile; Q2 = Second Quartile; Q3 = Third Quartile; Q4 = Fourth Quartile; SD = standard deviation.

Table 4. Descriptive Statistics and Distribution of Reporting-Arrival Interval (T2)

Descriptive Statistics

Location Type	N	Mean	SD	Q1	Q2	Q3	Q4
Overall	11071	7.055	4.919	4	6	9	104
Urban	6546	5.824	3.879	4	5	7	72
Rural	4525	8.836	5.664	5	8	11	104
Freeways or expressways	1846	8.246	4.687	5	7	10	65
Conventional roads	9225	6.816	4.930	4	6	8	104

Distribution

Required time	Overall	Urban	Rural	Freeway or expressway	Conventional road
5 min or less	30.4%	39.5%	17.3%	16.7%	33.2%
5–10 min	49.0%	49.7%	48.0%	52.8%	48.3%
11–15 min	14.4%	8.36%	23.1%	22.3%	12.8%
15–20 min	3.91%	1.50%	7.40%	5.85%	3.52%
21–25 min	1.29%	0.43%	2.54%	1.30%	1.29%
26–30 min	0.47%	0.23%	0.82%	0.60%	0.44%
More than 30 min	0.53%	0.28%	0.85%	0.45%	0.45%

Note: Q1 = First Quartile; Q2 = Second Quartile; Q3 = Third Quartile; Q4 = Fourth Quartile; SD = standard deviation.

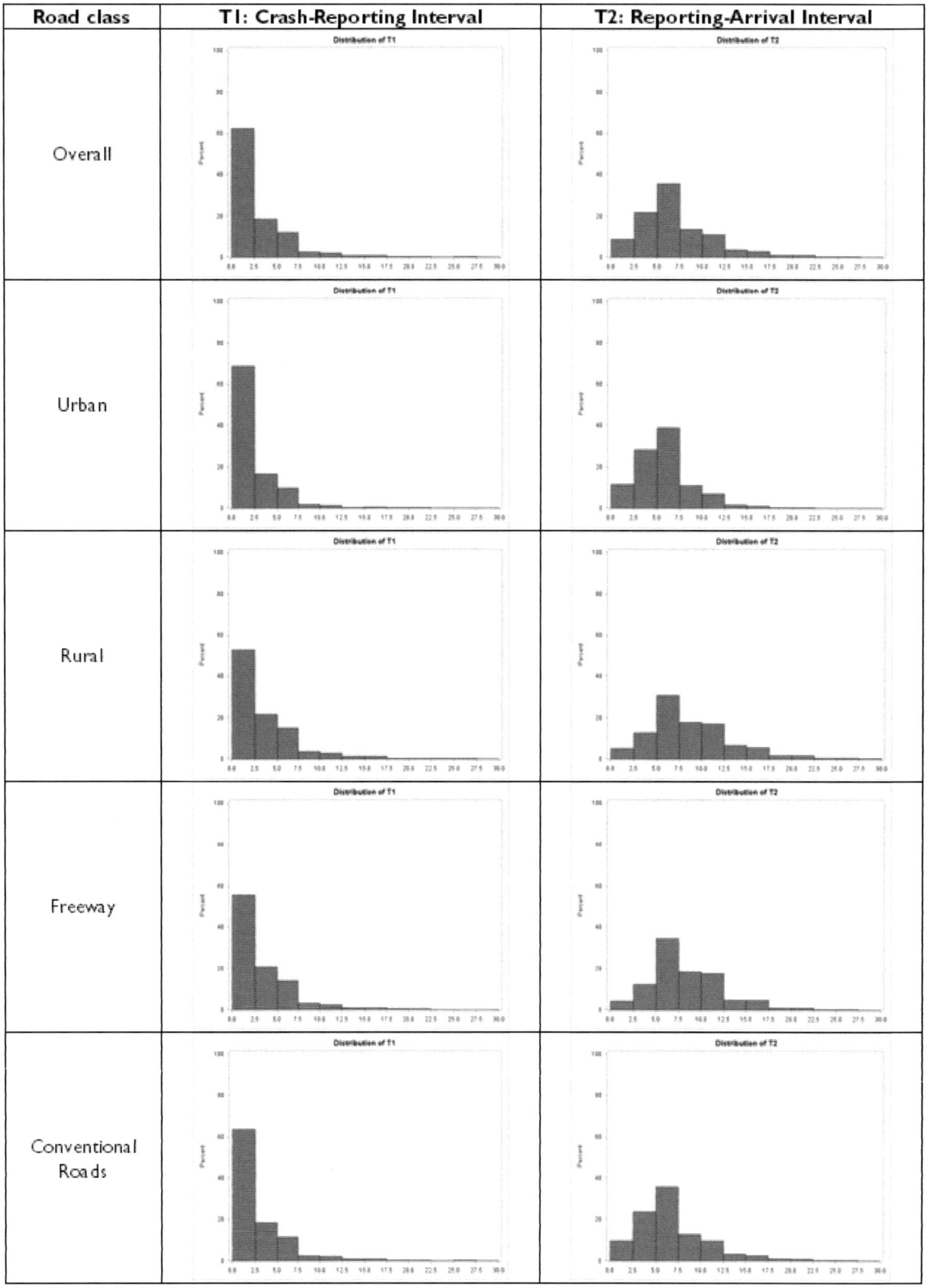

Figure 2. Histogram of crash-reporting (T1) and reporting-arrival (T2).

the EMS took more than 10 min to arrive at the scene. Nevertheless, there is a variance in crash-reporting and reporting-arrival intervals by geographic location (i.e., urban and rural) and roadway functional classification (i.e., freeway or expressway and conventional road). Thus, the research team analyzed crash-reporting and reporting-arrival intervals by location and roadway functional classification. It was shown that average crash-reporting intervals in the urban and rural areas are 3.03 and 4.53 min, respectively. About 73% of fatal crashes are reported within 5 min in the rural areas whereas almost 84% are reported within 5 min in the urban areas. In the rural areas, only 17% of EMS arrive on the scene within 5 min while 40% in the urban areas need 5 min or less for EMS arrival on the scene after the notification. In the case of average crash-reporting and reporting-arrival intervals by roadway functional classification, there is no considerable difference between the average crash-reporting interval on the freeway or expressway (3.82 min) and conventional road (3.61 min). In contrast, the average reporting-arrival interval on the freeways or expressways (8.25 min) is longer than that on the conventional roads (6.82 min).

Modeling Results

Table 5 summarizes the MAD values of the estimated models. It was found that log-logistic and gamma performed the best for crash-reporting (T1) and reporting-arrival intervals (T2), respectively. Still, the differences between the MAD values are very small. It implies that the other models perform equally well as the best models. For instance, the MAD of the log-normal model for T1 (2.886) is almost same as that of the log-logistic model (2.884). Similarly, the MAD of the log-logistic model for T2 (2.741) is nearly equal to that of the gamma model (2.740). Also, the significant variables were almost identical. Thus, it is possible that the best models in this study may be replaced if different dataset is used. Table 6 summarizes the modeling results for T1 and T2 with the smallest MAD. The modeling results are discussed in the following section.

Table 5. Comparison of MAD Values between the Models

Model	T1: Crash-reporting interval	T2: Reporting-scene arrival interval
Exponential	2.914	3.053
Gamma	2.894	**2.740**
Log-logistic	**2.884**	2.741
Log-normal	2.886	2.751
Weibull	2.918	2.743

Note: All other variables are significant at 95% confidence interval.

Discussion and Conclusions

The log-logistic model revealed many significant factors for T1 (Table 6). Firstly, 'Whether all crash-involved persons were killed' has a positive effect on T1, which indicates that T1 would be significantly prolonged if all crash-involved people were killed instantaneously, or severely injured. If any person survived from the crash and is conscious, this person can report the crash. On the other hand, it is also possible that the prolonged T1 resulted in deaths of all crash-involved persons. 'Log (vehicle-miles-traveled per square mile)' is the variable that shows the intensity of traffic volume in a unit area. It is noted that the traffic volume density is not of the roadway where the crash occurred. As the variable was processed from TAZs, it shows the traffic volume density of the zone. The traffic volume density variable has a negative coefficient, which implies that in areas with larger volume density a crash can be quickly detected and reported as there are more drivers and people. Along with the traffic volume density, there are several conditions in which a crash can be reported promptly such as 'Intersection', 'Urban (vs. rural)', 'Proportion of residential areas', 'Log (employments per square mile)', 'Log (school enrollments per square mile)', and 'Proportion of commuters who walk'. Furthermore, 'Hit-and-run involved' has a positive relationship with T1. According to several previous studies (*28–30*), hit-and-run is more frequent when the driver feels he or she can escape detection and there is a higher chance of leaving the crash scene quickly. Thus, the hit-and-run involved crashes are less likely to be reported immediately. Two functional classification variables, 'Freeway or expressway' and 'Arterial road' are positively associated with T1. It shows that crash occurrence on the roadway of higher classification (i.e., freeway or expressway, and arterial) may need more time to be reported compared to collectors and local roads. 'Curve' is positively correlated with T1. It may be because the crashes occurring on the curve section are less likely to be detectable than those on the straight section. Moreover, all dark conditions were found to have a positive association with T1. It is interesting to note that the dark condition without lighting has the largest effect, the dawn condition is the second largest, and the dark condition with lighting has the smallest effect. This may infer that it is possible to reduce the crash-reporting interval by installing more lighting at key locations.

Many significant factors were found from the gamma model for T2 as shown in Table 5. 'Whether all crash-involved persons were killed' was significant in the T2 model, as well as in the T1 model. There are two possibilities for a crash with no survivor: (1) all crash-involved

Table 6. Modeling Results

Category	Variable	Log-Logistic Model for T1			Gamma Model for T2		
		Estimate	Std. Err.	*p*-value	Estimate	Std. Err.	*p*-value
–	Intercept	1.4145	0.0388	<.0001	1.7844	0.0767	<.0001
Person	Whether all crash-involved persons were killed	0.0527	0.0180	0.0033	0.0728	0.0122	<.0001
Traffic	Log (vehicle-miles-traveled per square mile)	–0.0112	0.0046	0.0151	–	–	–
	Proportion of heavy vehicle traffic	–	–	–	1.1596	0.1181	<.0001
Functional classification	Freeway or expressway	0.1698	0.0238	<.0001	0.2299	0.0150	<.0001
	Arterial road	0.0376	0.0181	0.0380	–	–	–
	Collector	–	–	–	–	–	–
Roadway	Lane counts	–	–	–	–0.0192	0.0065	0.0032
	Curve	0.0845	0.0200	<.0001	–	–	–
	Log (road lengths per square mile)	–	–	–	–0.1035	0.0088	<.0001
Location	Intersection	–0.0861	0.0176	<.0001	–0.0748	0.0125	<.0001
	Interchange	–	–	–	–0.1103	0.0441	0.0124
	Urban (vs. rural)	–0.1457	0.0201	<.0001	–0.2190	0.0127	<.0001
Driver action	Hit-and-run involved	0.0835	0.0325	0.0102	–	–	–
Lighting condition	Dark without lighting	0.1970	0.0183	<.0001	0.0676	0.0122	<.0001
	Dark with lighting	0.0555	0.0190	0.0035	–	–	–
	Dawn	0.1348	0.0542	0.0129	0.1300	0.0378	0.0006
Land-use	Proportion of agriculture areas	–	–	–	0.1349	0.0252	<.0001
	Proportion of public areas	–	–	–	0.1546	0.0427	0.0003
	Proportion of residential areas	–0.1669	0.0363	<.0001	0.0837	0.0291	0.0040
Car-ownership	Proportion of families with 2 or more vehicles	–	–	–	0.1579	0.0356	<.0001
Socio-demographic	Log (employments per square mile)	–0.0444	0.0051	<.0001	–	–	–
	Log (school enrollments per square mile)	–0.0064	0.0029	0.0270	–0.0097	0.0021	<.0001
Commuter	Proportion of commuters using car	–	–	–	0.2046	0.0809	0.0114
	Proportion of commuters who walk	–0.7213	0.2893	0.0127	–	–	–
Log-logistic parameter	Scale	0.4412	0.0035	–	–	–	–
Gamma parameters	Scale	–	–	–	0.5451	0.0037	–
	Shape	–	–	–	0.2829	0.0173	–
MAD		2.884			2.740		

Note: – shows the variable was not statistically significant at 95% confidence level and excluded from the model. All other variables are significant at 95% confidence level.

persons were killed instantaneously at the time of crash and (2) there was a survivor at the time of crash who died after arrival of EMS personnel. For the first case, no occupants can report the crash and the crash must have been reported by another person (who was not involved in the crash), which will increase T1. T2 could be also increased as it would be difficult to find the exact location. For the second case, even if there was a survivor (who would die soon), it is very probable that he or she was seriously injured or unconscious and could have not reported the crash. It will result in an increase of T1 and T2. Several variables indicating that the location is in suburban or rural areas were found to have a positive relationship with T2, including 'Proportion of agriculture areas', 'Proportion of residential areas', 'Proportion of families with 2 or more vehicles', and 'Proportion of commuters using car'. The result showed that crashes happening in suburban or rural areas need more time for EMS arrival. On the other hand, there are three factors representing urban areas including 'Log (road lengths per square mile)', 'Urban (vs. rural)', and 'Log (school enrollments per square mile)'. Since these three variables have negative coefficients, it can be interpreted that the urban areas require shorter time for EMS to arrive. 'Freeway or expressway' has a positive sign, which implies more time would be required for EMS dispatch. Because freeways and expressways are full access-controlled roadways EMS vehicles can only access from ramps, which results in extended T2. Regarding the lighting condition, both 'Dark without lighting' and 'Dawn' is positively related to T2. It would be very hard to detect precise crash locations under such dark environments. It was also shown that shorter T2 is observed at 'Intersection' and 'Interchange'. It is possible that it is because these locations are easier to access. Lastly, 'Lane counts' has a negative relationship with T2. It suggests that multilane highways have shorter T2 compared to two-lane highways.

It is anticipated that the findings from this study can be used to develop effective and practical strategic plans to minimize EMS reporting and dispatch time and, therefore, decrease the possibility of fatalities. To reduce the crash-reporting interval (T1), the following policies or countermeasures may be suggested:

- Conduct frequent patrols on non-local roads with less traffic volume, particularly during the nighttime. Systems such as OnStar, and in the future connected vehicles technology, could lead to improvement of T1.
- Install emergency phones in the areas not covered by the cell phone network in rural areas.
- Police should pay more attention to the areas with low population, employment, and school enrollment for an early detection.
- Install more lighting at crash-prone locations.

The following policies or countermeasures may be effective to reduce the reporting-arrival (T2) interval:

- Retain sufficient EMS stations, equipment, and personnel, especially in rural areas.
- Install emergency access ramps for crash-prone locations on freeway or expressway, if economically and practically feasible.
- Use optimized paths to crash locations considering land-use characteristics along with real-time traffic information.
- Implement a signal pre-emption technology widely to allow EMS vehicles right-of-way to pass intersections more safely and promptly.

The strength of this study is that the authors have applied hazard-based duration models for EMS crash-reporting and reporting-arrival intervals, which has not been adopted in the literature. As a result, many factors have been revealed in this study. It was interesting that not only roadway and environmental conditions but also land-use and socio-demographic factors have significant effects on the promptness of EMS. Many previous safety studies commonly showed that land-use and socio-demographic factors are related to traffic safety (*31–39*); however, no study has attempted such factors for the EMS times. It is expected that the findings from this study can be used to develop effective and practical strategic plans to minimize EMS reporting and arrival time and, therefore, reduce permanent injuries and fatalities from traffic crashes. Another strength of this study is that data from multiple sources were used including the FARS, the Florida Department of Transportation, the U.S. Census Bureau, and the Florida Department of Revenue. Nevertheless, this study is not without limitations. For example, the authors were able to collect fatal crashes only. If non-fatal crash data are available, it is possible that the severity level of crashes has a significant influence on the EMS time. Furthermore, there are only three EMS-related variables that are provided from the FARS database: (1) Notification Time EMS; (2) Arrival Time EMS; and (3) EMS Arrival Time at Hospital. However, the third variable has too many missing values and the arrival-hospital duration could not be analyzed. It would be ideal if the EMS response level information were included such as lights, sirens, and so forth, but they were not available. Lastly, this study only used the FARS data. It would be an important extension if the data from vehicles equipped with an advanced technology like OnStar could be analyzed. These limitations need to be addressed and overcome in the future studies.

References

1. National Highway Traffic Safety Administration. *Traffic Safety Facts 2014*. DOT HS 812 262. National Center for Statistics and Analysis, Washington, DC, 2016.
2. Clark, D. E., J. Qian, K. C. Sihler, L. D. Hallagan, and R. A. Betensky. The Distribution of Survival Times after Injury. *World Journal of Surgery*, Vol. 36, No. 7, 2012, pp. 1562–1570.
3. Brodsky, H. The Call for Help after an Injury Road Accident. *Accident Analysis & Prevention*, Vol. 25, No. 2, 1993, pp. 123–130.
4. Sánchez-Mangas, R., A. García-Ferrrer, A. De Juan, and A. M. Arroyo. The Probability of Death in Road Traffic Accidents. How Important is a Quick Medical Response? *Accident Analysis & Prevention*, Vol. 42, No. 4, 2010, pp. 1048–1056.
5. Pell, J. P., J. M. Sirel, A. K. Marsden, I. Ford, and S. M. Cobbe. Effect of Reducing Ambulance Response Times on Deaths from out of Hospital Cardiac Arrest: Cohort Study. *British Medical Journal*, Vol. 322, No. 7299, 2001, pp. 1385–1388.
6. O'Keeffe, C., J. Nicholl, J. Turner, and S. Goodacre. Role of Ambulance Response Times in the Survival of Patients with Out-of-Hospital Cardiac Arrest. *Emergency Medicine Journal*, Vol. 28, No. 8, 2011, pp. 703–706.
7. Castrén, M., R. Karlsten, F. Lippert, E. F. Christensen, E. Bovim, A. Kvam, I. Robertson-Steel, J. Overton, T. Kraft, and L. Engerstrom. Recommended Guidelines for Reporting on Emergency Medical Dispatch When Conducting Research in Emergency Medicine: The Utstein Style. *Resuscitation*, Vol. 79, No. 2, 2008, pp. 193–197.
8. Yasunaga, H., H. Miyata, H. Horiguchi, S. Tanabe, M. Akahane, T. Ogawa, S. Koike, and T. Imamura. Population Density, Call-Response Interval, and Survival of Out-of-Hospital Cardiac Arrest. *International Journal of Health Geographics*, Vol. 10, No. 1, 2011, p. 1.
9. Evanco, W. M. The Potential Impact of Rural Mayday Systems on Vehicular Crash Fatalities. *Accident Analysis & Prevention*, Vol. 31, No. 5, 1999, pp. 455–462.

10. Gonzalez, R. P., G. Cummings, M. Mulekar, and C. B. Rodning. Increased Mortality in Rural Vehicular Trauma: Identifying Contributing Factors through Data Linkage. *Journal of Trauma and Acute Care Surgery*, Vol. 61, No. 2, 2006, pp. 404–409.
11. Brodsky, H. Delay in Ambulance Dispatch to Road Accidents. *American Journal of Public Health*, Vol. 82, No. 6, 1992, pp. 873–875.
12. Meng, Q, and J. Weng. Uncertainty Analysis of Accident Notification Time and Emergency Medical Service Response Time in Work Zone Traffic Accidents. Traffic Injury Prevention, Vol. 14, No. 2, 2013, pp. 150–158.
13. Yasmin, S., N. Eluru, and A. R. Pinjari. Analyzing the Continuum of Fatal Crashes: A Generalized Ordered Approach. *Analytic Methods in Accident Research*, Vol. 7, 2015, pp. 1–15.
14. Nam, D, and F. Mannering. An Exploratory Hazard-Based Analysis of Highway Incident Duration. *Transportation Research Part A: Policy and Practice*, Vol. 34, No. 2, 2000, pp. 85–102.
15. Chung, Y. Development of an Accident Duration Prediction Model on the Korean Freeway Systems. *Accident Analysis & Prevention*, Vol. 42, No. 1, 2010, pp. 282–289.
16. Bhat, C. R. A Generalized Multiple Durations Proportional Hazard Model with an Application to Activity Behavior During the Evening Work-to-Home Commute. *Transportation Research Part B: Methodological*, Vol. 30, No. 6, 1996, pp. 465–480.
17. Bhat, C. R., T. Frusti, H. Zhao, S. Schönfelder, and K. W. Axhausen. Intershopping Duration: An Analysis Using Multiweek Data. *Transportation Research Part B: Methodological*, Vol. 38, No. 1, 2004, pp. 39–60.
18. Lee, B, and H. J. Timmermans. A Latent Class Accelerated Hazard Model of Activity Episode Durations. *Transportation Research Part B: Methodological*, Vol. 41, No. 4, 2007, pp. 426–447.
19. van den Berg, P., T. Arentze, and H. Timmermans. A Latent Class Accelerated Hazard Model of Social Activity Duration. *Transportation Research Part A: Policy and Practice*, Vol. 46, No. 1, 2012, pp. 12–21.
20. Yamamoto, T., J.-L. Madre, and R. Kitamura. An Analysis of the Effects of French Vehicle Inspection Program and Grant for Scrappage on Household Vehicle Transaction. *Transportation Research Part B: Methodological*, Vol. 38, No. 10, 2004, pp. 905–926.
21. Chen, C, and D. Niemeier. A Mass Point Vehicle Scrappage Model. *Transportation Research Part B: Methodological*, Vol. 39, No. 5, 2005, pp. 401–415.
22. Chang, H.-L, and T.-H. Yeh. Exploratory Analysis of Motorcycle Holding Time Heterogeneity Using a Split-Population Duration Model. *Transportation Research Part A: Policy and Practice*, Vol. 41, No. 6, 2007, pp. 587–596.
23. Paselk, T. A, and F. L. Mannering. Use of Duration Models for Predicting Vehicular Delay at a US/Canadian Border Crossing. *Transportation*, Vol. 21, No. 3, 1994, pp. 249–270.
24. Guo, H., Z. Gao, X. Yang, X. Zhao, and W. Wang. Modeling Travel Time under the Influence of on-Street Parking. *Journal of Transportation Engineering*, Vol. 138, No. 2, 2011, pp. 229–235.
25. Yang, X., Z. Gao, H. Guo, and M. Huan. Survival Analysis of Car Travel Time near a Bus Stop in Developing Countries. *Science China Technological Sciences*, Vol. 55, No. 8, 2012, pp. 2355–2361.
26. Abdel-Aty, M., A. Oloufa, Y. Peng, T.-W. Shen, X. Yang, J. Lee, R. Copley, A. Ismail, F. Eady, R. Lalchan, and B. Jarvis. *Real Time Monitoring and Prediction of Reduced Visibility Events on Florida's Highways*. Project Report BDV24 962-01, Florida Department of Transportation, 2014.
27. Sharman, B., M. Roorda, and K. Habib. Comparison of Parametric and Nonparametric Hazard Models for Stop Durations on Urban Tours with Commercial Vehicles. *Transportation Research Record: Journal of the Transportation Research Board*, 2012. 2269: 117–126.
28. Solnick, S. J, and D. Hemenway. The Hit-and-Run in Fatal Pedestrian Accidents: Victims, Circumstances and Drivers. *Accident Analysis & Prevention*, Vol. 27, No. 5, 1995, pp. 643–649.
29. Tay, R., S. M. Rifaat, and H. C. Chin. A Logistic Model of the Effects of Roadway, Environmental, Vehicle, Crash and Driver Characteristics on Hit-and-Run Crashes. *Accident Analysis & Prevention*, Vol. 40, No. 4, 2008, pp. 1330–1336.
30. Tay, R., U. Barua, and L. Kattan. Factors Contributing to Hit-and-Run in Fatal Crashes. *Accident Analysis & Prevention*, Vol. 41, No. 2, 2009, pp. 227–233.
31. Abdel-Aty, M., J. Lee, C. Siddiqui, and K. Choi. Geographical Unit Based Analysis in the Context of Transportation Safety Planning. *Transportation Research Part A: Policy and Practice*, Vol. 49, 2013, pp. 62–75.
32. Lee, J., M. Abdel-Aty, and X. Jiang. Multivariate Crash Modeling for Motor Vehicle and Non-Motorized Modes at the Macroscopic Level. *Accident Analysis & Prevention*, Vol. 78, 2015, pp. 146–154.
33. Lee, J., M. Abdel-Aty, K. Choi, and H. Huang. Multi-Level Hot Zone Identification for Pedestrian Safety. *Accident Analysis & Prevention*, Vol. 76, 2015, pp. 64–73.
34. Lee, J., M. Abdel-Aty, and X. Jiang. Development of Zone System for Macro-Level Traffic Safety Analysis. *Journal of Transport Geography*, Vol. 38, 2014, pp. 13–21.
35. Cai, Q., J. Lee, N. Eluru, and M. Abdel-Aty. Macro-Level Pedestrian and Bicycle Crash Analysis: Incorporating Spatial Spillover Effects in Dual State Count Models. *Accident Analysis & Prevention*, Vol. 93, 2016, pp. 14–22.
36. Lee, J., M. Abdel-Aty, and K. Choi. Analysis of Residence Characteristics of at-Fault Drivers in Traffic Crashes. *Safety Science*, Vol. 68, 2014, pp. 6–13.
37. Dong, N., H. Huang, J. Lee, M. Gao, and M. Abdel-Aty. Macroscopic Hotspots Identification: A Bayesian Spatio-Temporal Interaction Approach. *Accident Analysis & Prevention*, Vol. 92, 2016, pp. 256–264.

38. Lee, J, and M. Abdel-Aty. Macro-Level Analysis of Bicycle Safety: Focusing on the Characteristics of Both Crash Location and Residence. *International Journal of Sustainable Transportation*, 2017, pp. 1–8.
39. Lee, J., S. Yasmin, N. Eluru, M. Abdel-Aty, and Q. Cai. Analysis of Crash Proportion by Vehicle Type at Traffic Analysis Zone Level: A Mixed Fractional Split Multinomial Logit Modeling Approach with Spatial Effects. *Accident Analysis & Prevention*, Vol. 111, 2018, pp. 12–22.

The Standing Committee on Transportation Safety Management (ANB10) peer-reviewed this paper (18-01821).

Article

TRR
JOURNAL OF THE TRANSPORTATION RESEARCH BOARD

Transportation Research Record
2018, Vol. 2672(32) 72–82

Article reuse guidelines:
sagepub.com/journals-permissions
DOI: 10.1177/0361198118788464
journals.sagepub.com/home/trr
SAGE

Vehicle Consumer Complaint Reports Involving Severe Incidents: Mining Large Contingency Tables

Subasish Das[1], Abhisek Mudgal[1], Anandi Dutta[2], and Srinivas R. Geedipally[1]

Abstract

According to 2010–2014 Fatality Analysis Reporting System (FARS) data, nearly 6.35% of fatal crashes happened as a result of vehicles' pre-existing manufacturing defects. The National Highway Traffic Safety Administration's (NHTSA) vehicle complaint database incorporates more than 1.37 million complaint reports (as of June 1, 2017). These reports contain extended information on vehicle-related disruptions. Around 5% of these reports involve some level of injury or fatalities. This study had two principal objectives, namely (1) perform knowledge discovery to understand the latent trends in consumer complaints, and (2) identify clusters with high relative reporting ratios from a large contingency table of vehicle models and associated complaints. To accomplish these objectives, 67,201 detailed reports associated with injury or fatalities from the NHTSA vehicle complaint database were examined. Exploratory text mining and empirical Bayes (EB) data mining were performed. Additionally, this study analyzed five years (2010–2014) of FARS data to examine the research findings. Results show that major vehicular defects are associated with air bags, brake systems, seat belts, and speed controls. The EB metrics identified several key '*vehicle model with major defect*' groups that require more attention. This study demonstrates the applicability of consumer complaints in identifying major vehicular defects as well as key groups of '*vehicle model with major defect*.' The findings of this study will provide a significant contribution to the reduction of crashes from vehicle-related disruptions. The research presented in this paper is crucial given the ongoing advancement of connected and automated vehicle technologies.

Traffic safety is a major concern because of the economic and social costs of traffic crashes. According to National Safety Council (NSC), more than 40,000 people died in traffic crashes in 2016, which is the highest count since 2007. This count is a 6% rise from 2015 and is up 14% from the 2014 figure. The estimated annual death rate is 1.25 deaths per 100 million vehicle miles traveled, an increase of 3% from 2015. The estimated cost of motor-vehicle deaths, injuries, and property damage in 2016 was $432.5 billion, an increase of 12% from 2015 (*1*). The costs include wage and productivity losses, medical cost, administrative prices, employer costs, and property damage costs. These statistics make Vision Zero goals difficult to attain. Countless research efforts on conventional crash data have been conducted to understand better the factors that influence the frequency and severity of crashes and to provide more effective safety-related countermeasures. However, the number of crashes is still at an unacceptable level, which is evident by the sharp rise in fatalities in the last two years. This shows that, in addition to current efforts, research needs to be conducted with additional resources and in newer directions. Newer sources of data (for example, naturalistic driving data, U.S. census demographic data, American Community Survey, National Highway Traffic Safety Administration [NHTSA] consumer complaints data) can fill the current research gap.

According to 2010–2014 Fatality Analysis Reporting System (FARS) data, in nearly 6.35% of fatal crashes, vehicles' pre-existing defects were involved (*2*). A limited number of studies were conducted to identify risk factors from vehicle defects. An in-depth analysis of the vehicle defect-related issues would help in understanding the association between vehicle condition and automotive safety. The NHTSA vehicle complaint database (*3*)

[1]Texas A&M Transportation Institute, Texas A&M University System, College Station, TX
[2]Computer Science and Engineering, Texas A&M University System, College Station, TX

Corresponding Author:
Address correspondence to Subasish Das: s-das@tti.tamu.edu

incorporates more than 1.37 million complaint reports (as of June 1, 2017). Around 5% of these reports involve some level of injury or fatalities. The complaints file contains all safety-related defect complaints received by NHTSA since January 1, 1995. This research sought to better understand the implications of the correlations present across injury outcomes in the vehicular complaint database. The current study design has two major goals: (1) conduct knowledge discovery to identify the latent trends from consumer complaints, and (2) detect clusters with high relative reporting ratios from a large contingency table of vehicle models and associated complaints. To accomplish the research goals, this study performed exploratory text mining and empirical Bayes (EB) data mining.

Earlier Work and Research Context

A limited number of studies exist where the relationship between vehicle defects and associated safety implications are studied. Moodley and Allopi performed a random survey of parked vehicles to investigate if the vehicles were adequately safe for driving (*4*). They conducted the observational study without touching the vehicles, and found that 24% of them had tire defects and 11% of them had defective lights. In order to devise methods to evaluate vehicles for safety it may be important to find if vehicle defects are contributing factors to crashes and, if so, which defects are most prominent. In the past, defects in the brake system or tires have been cited as the most frequent cause of crashes (*5*). Wolf found in 1962 that 6% of crashes in trucks were as a result of mechanical failure, and the Road Research Laboratory in the United Kingdom (UK) found that vehicle defects caused at least 18% of crashes (*6*). Researchers have found through crash reconstruction that vehicle defects contributed to 9% of the crashes (*7*).

More recently, Cuerden et al. investigated the effects of vehicle defects in traffic crashes (*8*). This study estimated that vehicle defects are likely to be a contributing factor in around 3% of crashes in the UK. The findings also showed that reducing the frequency of testing of newer vehicles is likely to have adverse road safety consequences. Schoor and Niekerk attempted to establish the contribution of mechanical failures in traffic crashes (*9*). Data obtained from accident response units indicate that tires and brakes were the main contributors to mechanical failures resulting in crashes. Barry et al. found that the air bag is effective for the survival of vehicle occupants in case of a crash. This implies that in the event of a crash, occupants in vehicles with a defective air bag system would have lesser protection and a higher chance of trauma. Barry et al. also found that the effectiveness of the air bag becomes less clear when seat belts are employed (*10*).

The NHTSA complaint data have great potential for identifying how a particular vehicle defect is related to crash outcome, as data regarding many instances of vehicle defects are available. With the NHTSA vehicle consumer complaint data, the frequency of the combination of vehicle defects and crash severity (no injury, less severe injury, or fatality) can be studied. In recent years, several studies have incorporated text mining in transportation engineering research: consumer complaint analysis (*11*, *12*), understanding public sentiments on safety enhancement and bike sharing (*13*, *14*), identifying research trends from transportation engineering conference papers and journals (*15–19*), crash narrative investigation (*20–23*), text analytics using social media mining (*24–27*), and teen driver survey (*28*). Ghazizadeh et al. used text mining and latent Dirichlet analysis to examine the hidden trends in the NHTSA complaint database. The analysis was done using data from January 1995 to June 2012 over a small sample of 2000 comments (*11*).

This paper is unique in several aspects: (1) the complete data set of vehicle defect complaints involving some level of injury or fatalities is used instead of a sample, (2) complaint data are contrasted with FARS data to ensure complaint data is not biased because of under-reporting, and (3) the analysis goes beyond that of Ghazizadeh et al. (*11*) by analyzing a large contingency table comprising vehicle defects and occurrence of crashes.

Methodology

To accomplish the research goals, the current study is divided into two major tasks: (1) perform descriptive statistics and text mining to determine latest trends in complaint texts and associated data, and (2) conduct EB data mining to identify key groups of "vehicle model with complaint" from a large contingency table with 67,201 complaint reports.

Descriptive Statistics and Text Mining

The data used in this research are based on consumer provided complaints collected by NHTSA. As of June 1, 2017, this database incorporates more than 1.37 million complaint reports in structured form with 33 variables. Around 5% of these reports involve some level of injury or fatalities. The complaints file contains all safety-related defect complaints received by NHTSA since January 1, 1995. Table 1 lists the descriptions of the final variables. The number of reports associated with some level of injury or deaths is 67,201 (around 4.9% of total complaint entries). However, approximately 25% of those complaint reports with injury or death are not associated with traffic crashes.

Table 1. Description of the Variables

Code	Description	Categories
MAKETXT[a]	Vehicle/equipment make	–
MODELTXT[b]	Vehicle/equipment model	–
YEARTXT[c]	Model year	9999 if unknown
CRASH	Was vehicle involved in a crash	"Yes" or "No"
INJURED	Number of persons injured	–
DEATHS	Number of fatalities	–
COMPDESC	Specific component's description	–
CDESCR	Description of the complaint	–
CMPL_TYPE[d]	Source of complaint code	CAG = consumer action group
		CON = forwarded from a congressional office
		DP = defect petition
		EVOQ = hotline VOQ
		EWR = early warning reporting
		INS = insurance company
		IVOQ = NHTSA web site
		LETR = consumer letter
		MAVQ = NHTSA mobile app
		MIVQ = NHTSA mobile app
		MVOQ = optical marked VOQ
		RC = recall complaint
		RP = recall petition
		SVOQ = portable safety complaint form (pdf)
		VOQ = NHTSA vehicle owners questionnaire

[a,b,c]Combined together for Vehicle Model for EB data mining.
[d]In the final dataset, the complaint code is divided into three broader groups: H_VOQ = hotline vehicle owner's questionnaire (VOQ) [original code: EVOQ]; NHTSA = NHTSA web site [original code: IVOQ]; other = other complaint codes [original codes: CAG, CON, DP, EWR, INS, LETR, MAVQ, MIVQ, MVOQ, RC, RP, SVOQ, VOQ].

Figure 1 shows key risk factors from three data sources: (1) dataset 1 [NHTSA: FI]: 1995–2017 NHTSA complaint data with severities (both crash and no crash incidents), (2) dataset 2 [NHTSA: FI + Crash]: 1995–2017 NHTSA complaint data with severities (only traffic crash incidents; subset of dataset 1), and (3) dataset 3 [FARS]: 2010–2014 FARS data. NHTSA provides two columns on complaints. "COMPDESC" provides short and specific problem-related complaints, and "CDESCR" provides a detailed description of the complaints. Risk factors from NHTSA were performed by the sentence stemming (identifying the key theme) method. In 2010–2014 FARS data (dataset 3), 224,319 vehicles were involved in fatal crashes. Of these, 14, 222 vehicles (around 6.35%) had vehicular defects. FARS uses a variable named "MFACTOR" to classify vehicle defects. NHTSA database shows that "*air bag*" is the most dominant attributor to crash. FARS does not include "*air bag*" in "MFACTOR" classification. It has separate coding for "air bag" deployment. FARS shows that around 38% of fatal crashes involve an "*air bag*" deployment issue, which is similar to what the NHTSA complaint data show. "Brake system" is the dominant attribute in all three data sources. "Tire/Wheels" is the second most significant factor in FARS crashes, which is also among the top five factors in dataset 1 and dataset 2. The top five risk factors ("*air bags*," "*brake system*," "*vehicle speed limit*," "*seat belts*," *and* "*tires/wheels*") contribute to 62% of all reports in dataset 1. Dataset 2 shows that top five risk factors ("*air bags*," "*brake system*," "*seat belts*," "*seat*," and "*vehicle speed limit*") are included in 77% of all reports. This is corroborated by Schoor and Niekerk, who found that defects in tires and brake system were the two most dominant factors that resulted in mechanical failure leading to a crash (*9*).

Dataset 1 and dataset 2 require further exploration for a clear understanding of the percentage distribution of the key categories. These two databases (involve severities irrespective of crash or no crash involvement) have five major contributing factors: airbag, seat/seat belts, braking system, speed control, and tires/wheels (as listed in Table 2). For airbag-related complaints, two major groups (airbags, and airbags: frontal) contribute nearly 95% of the complaints. Similarly, the top three categories (tires, tires: tread/belt, and wheels) comprise 95% of the tire or wheel-related complaints. Vehicle speed control and accelerator pedal are the dominating categories in the speed control variable. Hydraulic brake failures are the key contributing category in braking system complaints. Seats and seat beats contribute nearly 60% of the

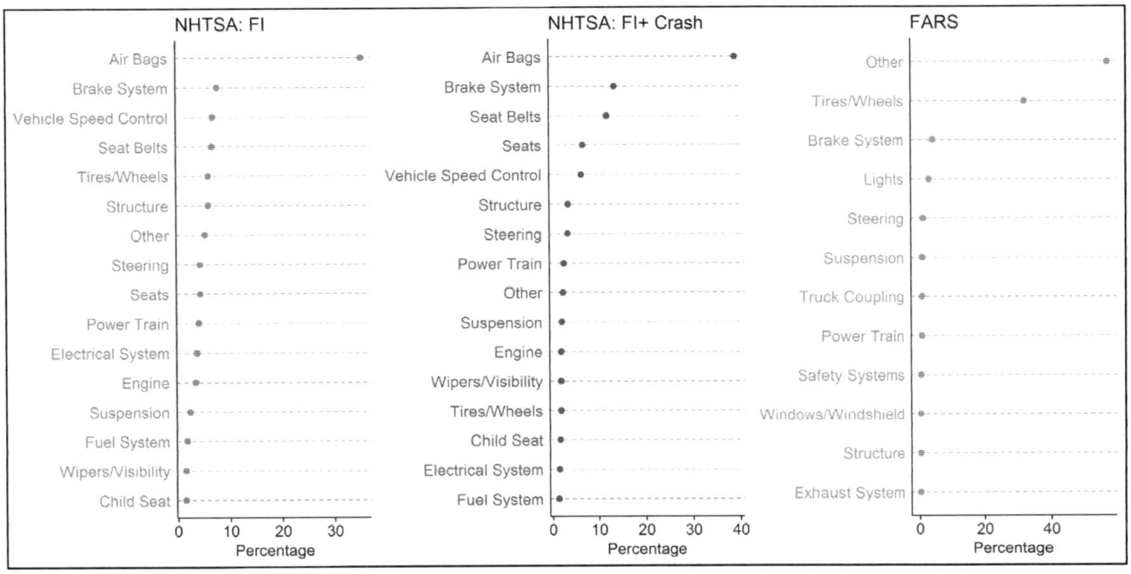

Figure 1. Comparison of major risk factors between NHTSA complaints and FARS.

Table 2. Top Categories under Major Complaints in NHTSA Compliant Databases

Complaint category	Percentage
Major complaint: airbag	
Air bags: frontal	47.2
Air bags	47.0
Air bags: side/window	4.7
Air bags: frontal: sensor/control module	0.7
Air bags: frontal: driver side inflator module	0.4
Major complaint: seat/seat belts	
Seat belts	31.9
Seats	28.2
Seats: front assembly: recliner	15.8
Seat belts: front: anchorage	13.2
Seat belts: front: retractor	10.8
Major complaint: tires/wheels	
Tires	52.3
Tires: tread/belt	30.4
Wheels	12.2
Tires: sidewall	3.5
Wheels: lugs/nuts/bolts	1.6
Major complaint: speed control	
Vehicle speed control	84.8
Vehicle speed control: accelerator pedal	8.8
Vehicle speed control: linkages	3.0
Vehicle speed control: cruise control	2.4
Vehicle speed control: cables	1.0
Major complaint: brakes	
Service brakes, hydraulic: antilock	41.5
Service brakes	25.9
Service brakes, hydraulic	18.9
Service brakes, hydraulic: foundation components	11.2
Parking brake	2.4

seat/seat belt complaint cases. Front seats and front seat belts comprise 40% of the seat/seat belt complaint cases. Future studies can focus on any of these major factors to identify more insights from the complaint database.

A comparison word cloud helps in studying the differences or similarities between two or more corpora by presenting the word cloud of each against the other (*29*). Consider that $p_{x,y}$ is the rate at which word x occurs in document y, and p_y is the average rate across documents $\left(\frac{\sum_y p_{x,y}}{n}\right)$, where n is the number of documents. In comparison clouds, the size of each word is mapped to its maximum deviation ($max_x(p_{x,y} - p_y)A = \pi r^2$) and its angular position is determined by the document in which that maximum occurs (*15*). For this analysis, the research team used complaint texts by broadly dividing them into two groups: crash, and no crash. Figure 2*a* shows comparison cloud for the corpora developed with vehicle models, and Figure 2*b* shows the comparison cloud for the corpora developed with vehicle complaints. The centrality of the word positioning indicates the presence of a centered word in both corpora. In Figure 2*a*, "*unknown*" and "*Toyota*" are dominant and central words in the "*No Crash*" group. This indicates that these two words are present in both groups, but the probability of presence in "*No Crash*" group is higher than in "*Crash*" group data (word frequencies were normalized before generating the word cloud).

In Figure 2*b*, dominant and central words are "*air*," "*bags*," "*speed*," "*service*," "*frontal*," and "*structure*." Five of these words fall in the "*No Crash*" group. Only

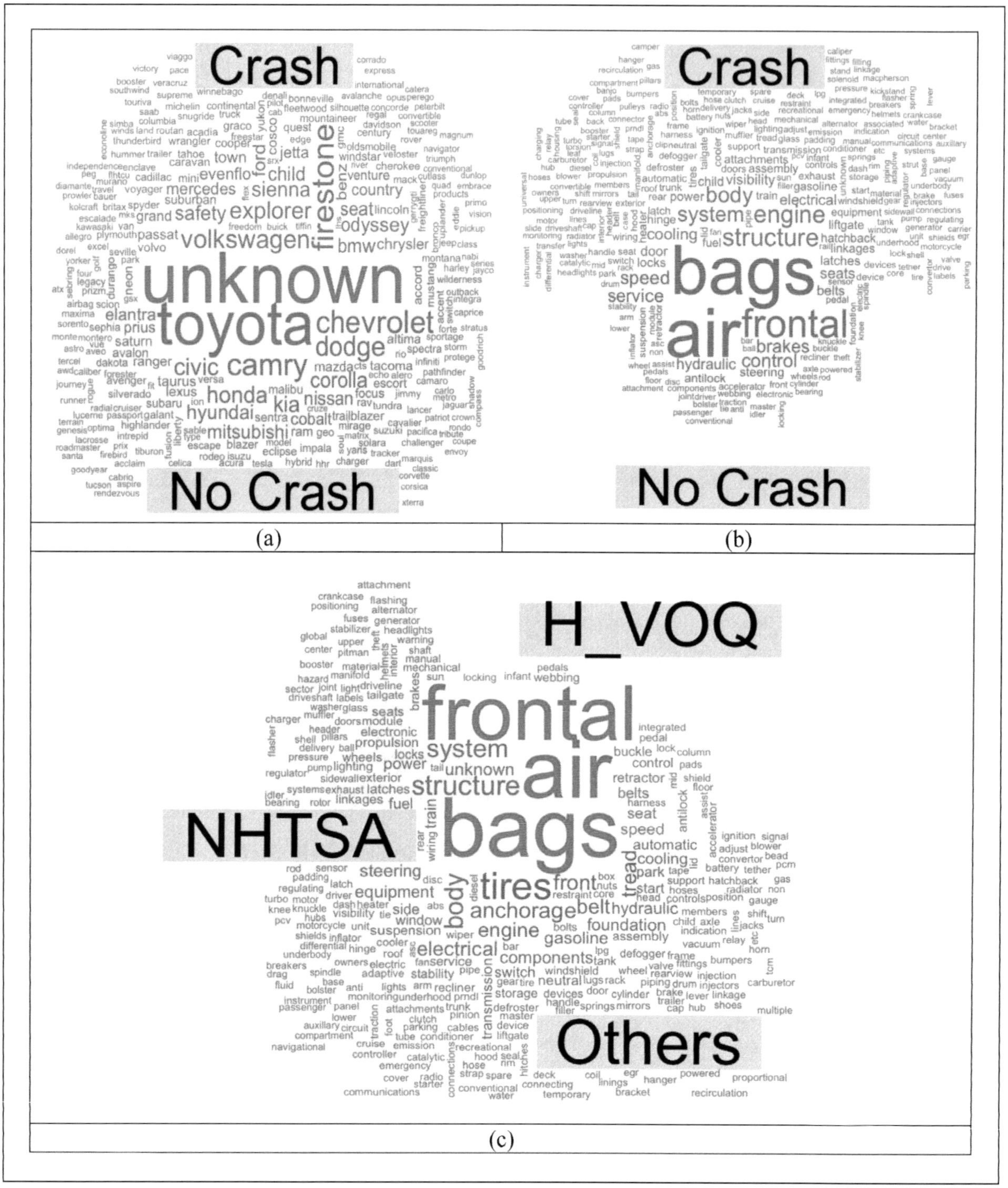

Figure 2. Comparison word clouds: (*a*) vehicle model; (*b*) customer complaints; (*c*) customer complaints clustered by complaint resources.

the word "structure" is inside the cloud boundary of "Crash"-related words. The findings indicate that the word group "*air bags*" is common in both corpora, but the probability of presence in "*No Crash*" is relatively higher. It also shows that body or structure, engine system, seat, and visibility are the dominant risk factors in

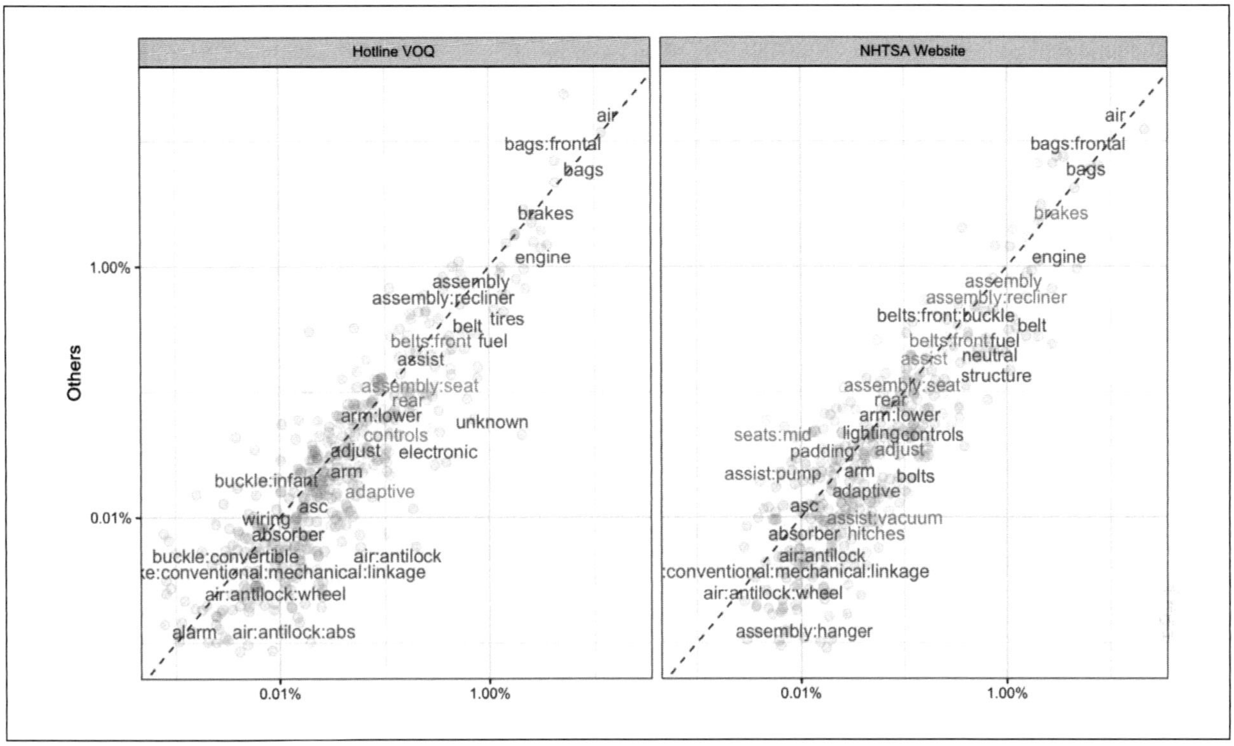

Figure 3. Comparing word frequencies of corpora based on complaint code.

the "Crash" group. Figure 2*c* shows comparison cloud for three groups based on the major complaint code. "*Air*," "*bag*," "*frontal*," "*tires*," "*structure*," "*belt*," "*body*," and "*electrical*" are the words with larger fonts. Most of these larger font words are in the center that represents their presence in all of the clusters.

Another approach is to examine the word frequencies of each corpus (as shown in Figure 3). Words that are close to the line in these plots have similar frequencies in both sets of texts, for example, in both "Hotline VOQ vs. Others" and "NHTSA website vs. Others" texts ("*air*," "*front bags*," "*bags*," "*brakes*" at the upper frequency end). Words that are far from the line are words that are found more in one set of texts than another is. For example, "*unknown*" word is a little far away from the center line in the "*Hotline VOQ vs. Others*" group (this word is not visible in "NHTSA website vs. Others" texts). The findings show that "Hotline VOQ" are associated with emergent defects like "*tire*," "*seat*," and "*engine*," and "*NHTSA website*" complaints are associated with the body or structural ("*structure*," "*bolts*," and "*lighting*") complaints.

Empirical Bayes Data Mining

The final NHTSA complaint database is a large contingency table. The association between complaints and vehicle model were analyzed using the EB geometric mean (EBGM) method, which is an association rules learning (ARM) method. ARM is a rule-based machine learning method that is used for discovering interesting connections between variables in a large frequency table. ARM was widely used in transportation safety research (*30–35*). Interested readers can consult for a short overview on the concepts of ARM.

The EBGM method helps to identify how significant is the frequency of an outcome given that it has a high occurrence. For example, it would help identify how significant is the occurrence of a certain defect in the vehicle given its model and model year. Complaint data may be affected by reporting bias, and therefore the association of complaints and vehicle model may not reflect the actual situation. However, the result of this analysis will create direction to further investigate the vehicle model and the occurrence of the given corresponding defects.

The analysis presented in this paper is performed using open source R package *openEBGM*, which estimates the significance of unusually large cell counts in large, sparse contingency table (*36*, *37*). This method can be used to find unusually high reporting rates of vehicle defects associated with the vehicle model.

Contingency tables can be analyzed using relative reporting ratio *RR* that is expressed by $^{N}/_{E}$, where N is the real count for the cell in the table and E is the expected frequency with an assumption of independency of rows and columns. However, *RR* tends to overstate the significance of a count when the real count is small. The *EBGM* method can be thought as a Bayesian approach to *RR* since *EBGM* adds Bayesian shrinkage corrections to *RR* such that high values that result from low *E* and low *N* are no longer identified as significant. Small *N* could occur by chance and N/E becomes high while it is not significant. The *EB* approach shrinks large *RR* toward 1 (the null hypothesis value) when *N* is small. The shrinkage is smaller for large counts and becomes almost negligible for very large counts. Therefore, the *EB* approach provides more stable results as compared with the *RR* measurement. The *EB* approach also accounts for sampling variation.

Theory. Consider a contingency table with cell count N_{ij} that follows a Poisson distribution with unknown mean μ_{ij}. Here *i* and *j* represent row and column numbers respectively. To evaluate cell counts N_{ij}, baseline or null hypothesis frequency, E_{ij} for each cell should be defined along with a statistical measure to rank cells according to their significance. A stratification variable, *k* is also defined to utilize the within group correlation. Consider the following group of equations defining E_{ij}:

$$N_{ij} = N_{ij\cdot} = \sum_k N_{ijk} \quad (1)$$

$$N_{i.j} = \sum_j N_{ijk} \quad (2)$$

$$N_{.jk} = \sum_i N_{ijk} \quad (3)$$

$$N_{..k} = \sum_i \sum_j N_{ijk} \quad (4)$$

$$E_{ij} = \sum_{jk} \frac{N_{i.k} N_{.jk}}{N_{..k}} \quad (5)$$

E_{ij} is the expected count assuming that the rows and columns variables are independent conditional on stratification variable. Consider $\lambda_{ij} = \frac{\mu_{ij}}{E_{ij}}$ as the decision statistics for evaluating unusually large frequencies for each cell in the contingency table. The statistic λ is drawn from a mixture of two gamma densities with (α_1, α_2) and (β_1, β_2) as the shape parameters and rate parameters and P as proportion constant of the two densities. Prior probability density for λ is given by $\pi(\lambda; \alpha_1, \beta_1, \alpha_2, \beta_2, P = P \times g(\lambda; \alpha_1, \beta_1) + (1 - P) \times g(\lambda; \alpha_2, \beta_2)$ where g is gamma distribution. The resultant density is a five parameter probability density function. Assuming that $E, \alpha_1, \beta_1, \alpha_2, \beta_2, and\ P$ are known, the marginal distribution of *N* follows a mixture of two negative binomial distributions. Consider $EBlog2_{ij}$, the posterior expectation of $log2(\lambda_{ij})$ as the Bayesian version of *RR*. It can be derived in the following form:

$$E[\log(\lambda)|N = n] = Q_n[\psi(\alpha_1 + n) - \log(\beta_1 + E)] + (1 - Q_n)[\psi(\alpha_2 + n) - \log(\beta_2 + E)] \quad (6)$$

where ψ is digamma function, the derivative of $\log[\Gamma(.)]$, and Q_n is the posterior probability that λ came from the first component of the mixture, given $N = n$.

For large values of *E* and N/E, $\psi(n)$ approaches $\log(n)$, then $EBlog2_{ij}$ or $EB[\log(\lambda)|N = n]$ approaches $\log(\alpha + N) - \log(\beta + N)$, *or* $\log\left(N/E\right)$ or $\log(RR)$. However, when *E* and N/E are not large, $EBlog2_{ij}$ would shrink toward a lower value. Finally, to make $EBlog2_{ij}$ of the same scale as *RR* and obtain a value that is easily comparable and interpretable, DuMuchel computes EBGM ($EBGM_{ij}$), the geometric mean of $EBlog2_{ij}$. It is given by the equation $EBGM_{ij} = 2^{EBlog2_{ij}}$. In EB methods, choice of prior parameters is obtained from the data themselves. In this methodology, $\alpha_1, \beta_1, \alpha_2, \beta_2,$ and *P* are obtained by maximizing the likelihood of these parameters taken together. This likelihood is the marginal distribution of N_{ij}.

Model and Estimates. A significantly large proportion of vehicle owners' complaints were identified using an R package *openEBGM*, which uses an EB data mining approach developed by DuMouchel. The EBGM method provides information about the significance of frequency of a given combination of effect and response in the contingency table. It works well even with very large but sparse tables. The final database contains 67,201 complaint records. There are 37,312 unique combinations of "Variable 1-Variable 2". Table 3 lists the variable used for the final analysis.

The variables are strata used for the modeling:

- Variable 1: Vehicle model (code developed by combining MAKETXT, MODELTXT, YEARTXT)
- Variable 2: Vehicle defect (COMPDESC)
- Strata 1: Crash involvement (CRASH)
- Strata 2: Complaint code (CMPL_TYPE)

The first step is to conduct actual counts of each vehicle model and defect combination, expected counts (*E*) under the row/column independence assumption, *RR*, and proportional reporting ratio (*PRR*). Stratification is beneficial in controlling confounding variables. For instance, complaints are affected by different complaint submission methods or whether the complaints were associated with crashes or not. Stratification will affect *E* and *RR*, but not *PRR*. The *E*s are calculated by summing the expected counts from every stratum. Ideally, each stratum should contain several unique reports to ensure

Table 3. Top 10 Combination Groups with High PRRs

Vehicle model/component model	Complaint	E	RR	PRR	EBGM
Accubuilt-Limousine (2001)	Wheels: rim	0.0018	1110.1	1080.94	1.16
Aai Motorsports-Combination Lamps (unknown)	Equipment: motorcycle: helmets	0.0020	1009.18	1015.44	1.16
Acura-Acura (1999)	Seats: mid/rear assembly	0.0005	2018.36	957.44	1.06
A-1 Alternative Fuel Syst-Cng Fuel System (unknown)	Equipment: electrical	0.0043	231.46	744.68	1.06
Ace Electric-Ace (unknown)	Equipment: electrical	0.0043	231.46	744.68	1.06
Acura-Acura (1996)	Suspension: front: control arm: lower ball joint	0.0015	648.17	609.28	1.06
Ac Delco-18M615 (unknown)	Electrical system: battery	0.0029	347.19	478.71	1.06
Accessory-Helmet (unknown)	Equipment: other: labels	0.0046	217.95	462.19	1.06
Accubuilt-Limousine (2001)	Parking brake	0.0067	300.03	331.77	1.16
Ac Delco-45A3073 (unknown)	Equipment: mechanical	0.0073	137.05	272.43	1.06

good estimates of E. The actual counts (N) and expected counts (E) are used to estimate the hyperparameters of the prior distribution. A large contingency table will have many cells, resulting in computational difficulties for the optimization routines needed for estimation. Data squashing transforms a set of 2-dimensional points to a smaller number of 3-dimensional points. The idea is to reduce a large number of points (N, E) to a smaller number of points (N_k, E_k, W_k), where $k = 1, \ldots, M$ and W_k is the weight of the kth "superpoint". To minimize information loss, only points close to each other should be squashed. To determine the hyper parameters efficiently, around 10% of the reports were used for squashed data. The hyper parameters are estimated by minimizing the negative log-likelihood function. The optimized hyper parameters are $(\alpha_1, \beta_1, \alpha_2, \beta_2, P) = (2.32 \times 10^{-08}, 0.1052, 10.0441, 9.9493, 0.3362)$. EBGM is a measure of central tendency of the posterior distributions $\lambda_{ij}|N = n$. Scores much larger than 1 indicate vehicle model and complaint event pairs that are reported at an unusually high rate. Table 3 lists the top 10 combination groups with high PRRs.

EBGM is the antilog of the mean of the *ln*-transformed posterior distribution. It can be used as a measure of central tendency of the posterior distribution. The EBGM scores indicate adjusted estimate for the relative reporting ratio. For example, "*Ford-Explorer (2002) - Structure: Body: Hatchback/Liftgate: Hinge and attachments*" pair has an EB score of 55.14. The interpretation is that this pair occurs in the data 55.14 times more frequently than expected under the assumption of no association between the vehicle model and complaint. The 5% and 95% quantiles of the posterior distributions can be used to create two-sided 90% credibility intervals for λ_{ij}, given N_{ij}. As a result of the Bayesian shrinkage property, the EB scores are much more stable than *RR* for small counts. Table 4 lists the top 20 combination groups with high EBGM and quantiles (the list is sorted in descending order based on quantile 5% values). Figure 4 illustrates the values of Table 4 for easy interpretation.

A closer look at these vehicle model–complaint combinations can be helpful in identifying combination groups that may need further investigation. It is important to note that the EB scores find statistical associations, not causal relationships of any kind. Future investigations are required before a solid quantification of the scale of adverse effects of these combination groups can be performed. It is interesting that except one recent vehicle, all of the vehicles are older vehicles. As the new vehicles have been available to people for fewer years, these vehicle models would usually not show up in the top 20 groups. The authors envision using this database for future exploration to investigate whether there are other plausible reasons.

Conclusion

All vehicles depreciate in performance over time, which can have an adverse effect upon safety and roadway environment. Crashes are random incidents, and crashes associated with vehicle defects are more infrequent. Conventional crash reports do not include vehicle defects in detail. The NHTSA complaint database can be helpful in filling the current knowledge gap in identifying vehicle defects associated with crashes. The paper presented an analysis of a high-level evaluation of risk factors associated with vehicle defects. This analysis is based on the part of NHTSA consumer complaint database that includes records involving either injury or deaths. Some of the findings were compared with the FARS database.

The study has concluded that:

- The vehicle defects are likely to be a contributing factor for severe crashes for nearly 5% of the complaint database. For FARS, the rate is slightly higher. Vehicle defects contribute to 6.35% of all fatalities in FARS.

Table 4. Problem Groups with High EBGM Scores and Quantiles

Vehicle model/component model	Complaint	EBGM	Quantile 5%	Quantile 95%
Ford-Explorer (2002)	Structure: body: hatchback/liftgate: hinge and attachments	55.14	41.69	71.82
Jeep-Grand Cherokee (1998)	Power train: automatic transmission: park/neutral start switch	44.38	36.54	53.48
Jeep-Grand Cherokee (1999)	Power train: automatic transmission: park/neutral start switch	40.13	27.8	56.43
Jeep-Grand Cherokee (1996)	Power train: automatic transmission: park/neutral start switch	34.01	26.26	43.47
Jeep-Grand Cherokee (2001)	Power train: automatic transmission: park/neutral start switch	39.94	26.22	58.81
Jeep-Liberty (2002)	Suspension: front: control arm: lower ball joint	40.41	26.16	60.21
Hyundai-Veloster (2012)	Visibility: sun roof assembly	44.53	25.4	73.86
Jeep-Grand Cherokee (1997)	Power train: automatic transmission: park/neutral start switch	32.34	24.14	42.59
Volkswagen-Jetta (2002)	Seats: front assembly: seat heater/cooler	29.16	21.77	38.41
Pontiac-Trans Sport (1996)	Engine and engine cooling	38.14	21.69	63.94
Volkswagen-Passat (2003)	Seats: front assembly: seat heater/cooler	30.24	21.45	41.66
Volkswagen-Jetta (2003)	Seats: front assembly: seat heater/cooler	27.09	20.03	35.98
Chevrolet-S10 (2000)	Structure: body: tailgate: hinge and attachments	35.2	19.9	59.75
Toyota-Sienna (2004)	Structure: body: hatchback/liftgate: support device	25.69	18.35	35.18
Firestone-ATX (unknown)	Tires: tread/belt	21.7	16.64	27.89
Firestone-Wilderness (unknown)	Tires: tread/belt	20.91	16.29	26.52
Chevrolet-Suburban (1994)	Service brakes, hydraulic: antilock	22.08	15.87	30.05
GMC-Jimmy (1994)	Service brakes, hydraulic: antilock	15.78	23.46	33.85
Toyota-Sienna (2004)	Latches/locks/linkages: hatchback/liftgate: lock	15.7	25.04	38.39
Jeep-Grand Cherokee (1995)	Power train: automatic transmission: park/neutral start switch	15.53	22.43	31.54

- Both NHTSA and FARS show that air bags were deployed significantly (around 35%) in severe crashes.
- Tire/wheel was found as a key contributing factor in FARS data, but not in NHTSA complaints. The findings of FARS are in line with Schoor and Niekerk's study (*9*).
- The comparison cloud for vehicle models shows that some vehicle models were overrepresented in the "crash" group, although this study implies an association, not causation.
- According to NHTSA complaints, the key factors of vehicular defects include "*air bags*," "*brake system*," "*seat belts*," and "*speed control*." "*Seat belt*" and "*brake system*" were not found as key contributing factors in FARS. "*Brake system*" was found to be a key contributing factor in Schoor and Niekerk's study (*9*). These findings require further investigations.
- The NHTSA complaint database forms a large contingency table with 37,312 unique combination group records associated with vehicle model and complaint type. Simple count analysis will not consider the effect of other strata and additional local and global effects. EBGM is an appropriate tool for analyzing such database. By using EBGM, this study identifies the top 20 combination groups that were more likely to be associated with higher risks.

The findings from the NHTSA complaint data are based on reporting rates and not occurrence rates of a certain combination of vehicle and defects. This is because, unlike FARS data, NHTSA complaint data can suffer from under-reporting. Nevertheless, since the summary parameters (such as key defects associated with crashes and percentage of severe crashes involving vehicle defects) from both NHTSA complaint data and FARS data are comparable, the NHTSA data can be assumed to be a representative sample of the population of vehicle defects and crash data. Consequently, the results obtained from EB data mining provide a significant contribution to the area of safety in reducing crashes from vehicular manufacturing defects. Future studies with different subset data from NHTSA complaints (for example, tire or fuel system complaints; vehicle service life; restraint) could add value to the related research areas. With the advancement of connected and autonomous vehicle technologies, research such as that presented in this paper might be critical in maximizing the benefits of these technologies while minimizing system malfunctioning and associated injuries and fatalities.

This study aimed to play the role of a starting point in conducting research by performing both text mining and EBGM in analyzing complex datasets like the NHTSA vehicle consumer complaint database. This study can open doors for future studies on this database, which has

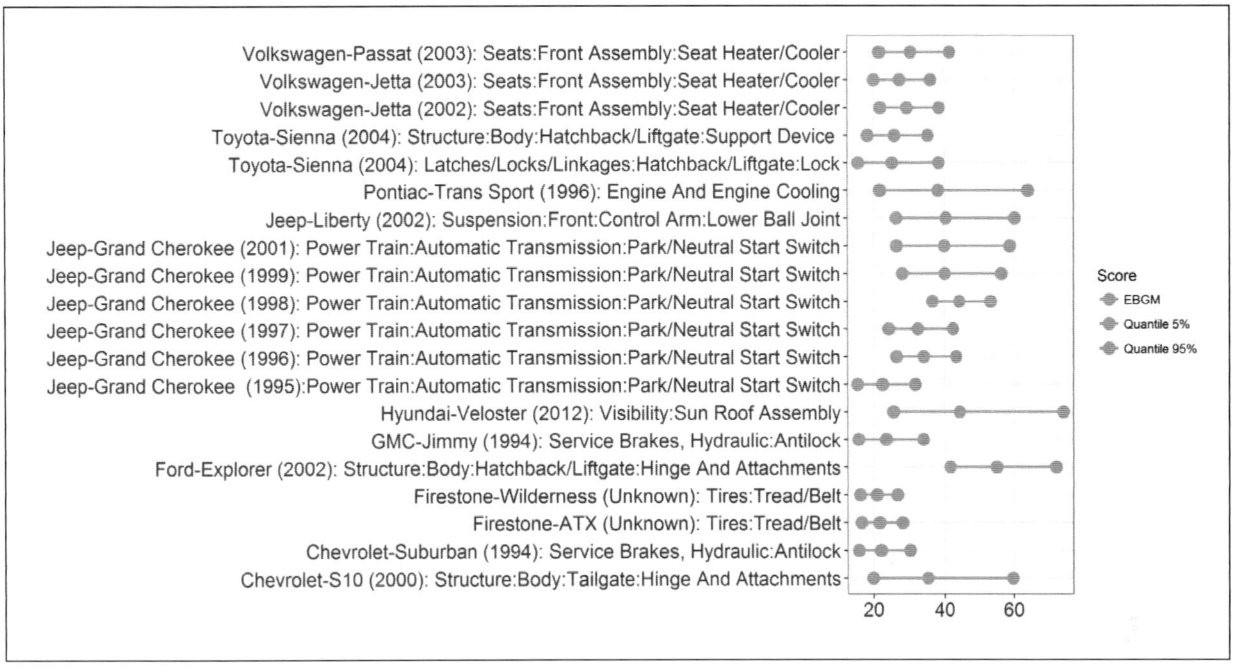

Figure 4. Vehicle/component model and complaint groups with higher EBGM.

not been explored in depth in earlier studies. The current study has several limitations. This study has not used the complete database of NHTSA vehicle consumer complaints; rather, it used complaint cases that involve at least one injury. An extended analysis on the complete database would identify the nature and distribution of various complaint categories to identify key patterns for possible solutions. Another limitation is that the database was compared with FARS that involves only fatal crashes. It would be more intuitive to compare this database with General Estimates System (GES) data of the National Automotive Sampling System, as GES represents national estimates for all levels of severities. Moreover, extended analysis on vehicle/component model and complaint groups would identify more vehicle models and associated major complaints. Current limitations in this study can be addressed in future studies.

Acknowledgments

The authors appreciate the assistance provided by Mr. Shashank Mehrotra.

Author Contributions

The authors confirm contribution to the paper as follows: study conception and design: Subasish Das; data collection: Subasish Das; analysis and interpretation of results: Subasish Das, Anandi Dutta, Abhisek Mudgal; draft manuscript preparation: Subasish Das, Anandi Dutta, Abhisek Mudgal, and Srinivas Geedipally. All authors reviewed the results and approved the final version of the manuscript.

References

1. *NSC Motor Vehicle Fatality Estimates- 2016*, 2015.
2. National Highway Traffic Safety Administration. *Fatality Analysis and Reporting System*. ftp://ftp.nhtsa.dot.gov/fars/. Accessed July 31, 2017.
3. National Highway Traffic Safety Administration. *Office of Defects Investigation*. http://www-odi.nhtsa.dot.gov/downloads/. Accessed July 31, 2017.
4. Moodley, S., and D. Allopi. An Analytical Study of Vehicle Defects and Their Contribution to Road Accidents. *Proc., 27th Southern Africa Transport Conference*, Pretoria, South Africa, 2008.
5. U.S. Department of Transportation. *National Traffic Safety Newsletter*, NHTSA Washington D.C., 1975.
6. Wolf, R. *Truck Accidents and Traffic Safety – An Overview*. SAE Meeting, Detroit, 1968.
7. Kinsey, G. Contribution of Unroadworthy Vehicles to Accidents. *Proc., 2nd Conference*, National Institute of Transport and Road Research, Pieterburg, South Africa, 1976.
8. Cuerden, R., M. Edwards, and M. Pittman. Effect of Vehicle Defects in Road Accidents. *Transport Research Laboratory Report No. PPR565*, TRL, Berkshire, UK, 2011.
9. Schoor, O. V., and J. L. Niekerk. Mechanical Failures as a Contributing Cause to Motor Vehicle Accidents-South Africa. *Accident Analysis & Prevention*, Vol. 33, No. 6, 2001, pp. 713–721.
10. Barry, S., S. Ginpil, and T. J. O'Neill, The Effectiveness of Air Bags. *Accident Analysis & Prevention*, Vol. 31, No. 6, 1999, pp. 781–787.

11. Ghazizadeh, M., A. McDonald, and J. Lee. Text Mining to Decipher Free-Response Consumer Complaints: Insights from the NHTSA Vehicle Owner's Complaint Database. *Human Factors: The Journal of the Human Factors and Ergonomics Society*, Vol. 56, No. 6, 2014, pp. 1189–1203.
12. Mehrotra, S., and S. Roberts. Identification and Validation of Themes from Vehicle Owner Complaints and Fatality Reports using Text Analysis. Presented at 97th Annual Meeting of the Transportation Research Board, Washington, D.C., 2018.
13. Das, S., X. Sun, and A. Dutta. Investigating User Ridership Sentiments for Bike Sharing Programs. *Journal of Transportation Technologies*, Vol. 5, 2015, pp. 69–75.
14. Chen, F., and R. Krishnan. *Transportation Sentiment Analysis for Safety Enhancement*. Report: DTRT12GUTG11. 2013.
15. Das, S., X. Sun, and A. Dutta. Text Mining and Topic Modeling of Compendiums of Papers from Transportation Research Board Annual Meetings. *Transportation Research Record: Journal of the Transportation Research Board*, 2016. 2552: 48–56.
16. Das, S., K. Dixon, X. Sun, A. Dutta, and M. Zupancich. Trends in Transportation Research: Exploring Content Analysis in Topics. *Transportation Research Record: Journal of the Transportation Research Board*, 2017. 2552: 27–38.
17. Boyer, R., W. Scherer, and M. Smith. Trends Over Two Decades of Transportation Research: A Machine Learning Approach. *Transportation Research Record: Journal of the Transportation Research Board*, 2017. 2552: 27–38.
18. Sun, L., and Y. Yin. Discovering Themes and Trends in Transportation Research Using Topic Modeling. *Transportation Research Part C: Emerging Technologies*, Vol. 77, 2017, pp. 49–66.
19. Das, S., and A. Dutta. Knowledge Extraction from Transportation Research Thesaurus. Presented at 97th Annual Meeting of the Transportation Research Board, Washington, D.C., 2018.
20. Das, S., A. Dutta, and M. Zupancich. Text mining on 100 years of Air Crash Narratives: Key Findings. Presented at 96th Annual Meeting of the Transportation Research Board, Washington, D.C., 2017.
21. Das, S., B. Brimley, T. Lindheimer, and M. Zupancich. Association of Reduced Visibility with Crash Outcomes. *IATSS Research*, 2017. https://doi.org/10.1016/j.iatssr.2017.10.003
22. Brown, D. Text Mining the Contributors to Rail Accidents. *IEEE Transactions on Intelligent Transportation Systems*, Vol. 17, 2016, pp. 346–355.
23. Bazargan, M., M. Johnson, and A. Vijayanarayanan. *An Evaluation of AIRES and STATISTICA Text Mining Tools as Applied to General Aviation Accidents*. Report: DOT/FAA/TC-TN13/7, FHWA, Washington, D.C., 2013.
24. Das, S., X. Sun, M. Zupancich, and A. Dutta. Twitter in Circulating Transportation Information: A Case Study on Two Cities. Presented at 96th Annual Meeting of the Transportation Research Board, Washington, D.C., 2017.
25. Gu, Y., Z. Qian, and F. Chen. From Twitter to Detector: Real-time Traffic Incident Detection Using Social Media Data. *Transportation Research Part C: Emerging Technologies*, Vol. 67, 2016, pp. 321–342.
26. Das, S., L. Minjares-Kyle, K. Dixon, A. Palanisamy, and A. Dutta. #TRBAM: Understanding Communication Patterns and Research Trends by Twitter Mining. Presented at 97th Annual Meeting of the Transportation Research Board, Washington, D.C., 2018.
27. Das, S., G. Medina, L. Minjares-Kyle, and Z. Elgart. Social Media Hashtags Associated with Bike Commuting: Applying Natural Language Processing Tools. Presented at 97th Annual Meeting of the Transportation Research Board, Washington, D.C., 2018.
28. Minjares-Kyle, L., S. Das, G. Medina, and R. Henk. Knowledge about Crash Risk Factors and Self-Reported Driving Behavior: Exploratory Analysis on Multi-State Teen Driver Survey. Presented at 97th Annual Meeting of the Transportation Research Board, Washington, D.C., 2018.
29. Gohil, A. *R Data Visualization Cookbook*. Packt Publishing, 2015.
30. Geurts, K., G. Wets, T. Brijs, and K. Vanhoof Profiling of High-Frequency Accident Locations by Use of Association Rules. *Transportation Research Record: Journal of the Transportation Research Board*, 2003. 1840: 123–130.
31. Pande, A., and M. Abdel-Aty. Discovering Indirect Associations in Crash Data Through Probe Attributes. *Transportation Research Record: Journal of the Transportation Research Board*, 2008. 2083: 170–179.
32. Das, S., and X. Sun. Investigating the Pattern of Traffic Crashes under Rainy Weather by Association Rules in Data Mining. Presented at 93rd Annual Meeting of the Transportation Research Board, Washington, D.C., 2014.
33. Das, S., L. Minjares-Kyle, R. Avelar, K. Dixon, and B. Bommanayakanahalli. Improper Passing-Related Crashes on Rural Roadways: Using Association Rules Negative Binomial Miner. Presented at 96th Annual Meeting of the Transportation Research Board, Washington, D.C., 2017.
34. Das, S., A. Dutta, R. Avelar, K. Dixon, X. Sun, and M. Jalayer. Supervised Association Rules Mining on Pedestrian Crashes in Urban Areas: Identifying Patterns for Appropriate Countermeasures. *International Journal of Urban Sciences*, 2018. https://doi.org/10.1080/12265934.2018.1431146
35. Das, S., A. Dutta, M. Jalayer, A. Bibeka, and L. Wu. Factors influencing the Patterns of Wrong Way Driving Crashes on Freeway Exit Ramps and Median Crossovers: Exploration using 'Eclat' Association Rules to Promote Safety. *International Journal of Transportation Science and Technology*, Vol. 7, No. 2, 2018, pp. 114–123. https://doi.org/10.1016/j.ijtst.2018.02.001.
36. Ihrie, J, I. Ahmed, and A. Poncet. *EBGM Scores for Mining Large Contingency Tables. Version 0.1.0*, 2017.
37. The R Foundation for Statistical Computing. *R: A Language and Environment for Statistical Computing, Version 3.4.1*, 2017.

The Standing Committee on Transportation Safety Management (ANB10) peer-reviewed this paper (18-03567).

Article

TRR
JOURNAL OF THE TRANSPORTATION RESEARCH BOARD

Transportation Research Record
2018, Vol. 2672(32) 83–93

DOI: 10.1177/0361198118791363
journals.sagepub.com/home/trr
SAGE

Service Location Optimization Model for Improving Rural Emergency Medical Services

Zhaoxiang He[1], Xiao Qin[1], Yuanchang Xie[2], and Jianhua Guo[3]

Abstract

Approximately 35,000 fatalities are attributed to accidents on U.S. highways each year and more than half of them occurred in rural areas. With such a high percentage of fatalities, rural areas are in critical need of timely and reliable Emergency Medical Services (EMS). EMS provide important prehospital care to victims before they are transferred to a hospital. After an accident occurs, the time it takes for victims to receive care from EMS is crucial to their survival. Compared with urban EMS, rural EMS face multiple challenges. One of them is how to properly site EMS stations to provide cost-effective services in rural areas. The goals of this paper include analyzing the spatial patterns of EMS station and incident locations, and optimizing rural EMS station locations. The data were collected from South Dakota, a rural state. This dataset was used to perform spatial analysis and to develop and evaluate an EMS location optimization model. The location optimization model aims to maximize the rural EMS coverage while taking service equity into consideration. The model was solved by a genetic algorithm toolbox in R. The proposed model provides an important and practical tool for rural EMS officials to select new EMS stations or relocate existing stations to improve service performance under budget and resource constraints.

In 2015 alone, more than 35,000 people lost their lives on U.S. highways. More than half of these fatalities occurred in rural areas. The fatality rate, defined as the rate of crash-related deaths per 100 million miles traveled, in rural areas is 2.6 times higher than in urban areas (*1*). Much of the difference in the fatality rate between urban and rural areas can be attributed to the increased travel time needed to reach a victim in rural areas (*1*). Because rural areas have a sparsely distributed population and road network, and a limited number of Emergency Medical Services (EMS) stations, long travel distances are often expected for most EMS service trips. Obviously, a major challenge here is how to provide timely service in these time-sensitive situations. It is important to develop methods to decrease the travel time needed for EMS to reach a victim as this time reduction may significantly increase the victim's chances of survival. The National Cooperative Highway Research Program (NCHRP) 500 report entitled *Volume 15: A Guide for Enhancing Rural Emergency Medical Services* puts special emphasis on reducing "the time from injury to appropriate definitive care" to improve rural EMS (*2*). In addition, this report identifies some challenges related to rural EMS, including geographic barriers, lack of professionals/paraprofessionals, aging or inadequate equipment, the absence of specialized EMS care, and local medical facilities (*3*).

With the objective of improving rural EMS performance, this study focuses on the locations of EMS stations. The goals of this study are to analyze the spatial patterns of EMS station and incident locations, and to optimize rural EMS station locations. Based on the spatial pattern analysis using EMS data from South Dakota, recommendations are provided to optimally locate EMS sites to obtain improved service performance. In addition, an EMS location optimization model is proposed. A case study using this proposed model and the South Dakota data is presented.

[1]Department of Civil and Environmental Engineering, University of Wisconsin-Milwaukee, Milwaukee, WI
[2]Department of Civil and Environmental Engineering, University of Massachusetts Lowell, MA
[3]Intelligent Transportation System Research Center, Southeast University, Nanjing, Jiangsu, PR China

Corresponding Author:
Address correspondence to Xiao Qin: qinx@uwm.edu

Literature Review

The National Highway Traffic Safety Administration (NHTSA) developed 35 performance measures for local EMS systems, including time-based variables (e.g., Mean Emergency Patient Response Interval) and outcome-based variables (e.g., EMS Cardiac Arrest Survival Rate to Emergency Department Discharge) (*2*). To evaluate EMS performance, other metrics were also proposed, such as time-related, distance-related, and outcome-based variables. Among these performance indicators, response time is considered as a major performance index and has been extensively used to evaluate EMS performance (*2*, *4*, *5*). Although a patient's treatment outcome depends on many factors such as the injury severity and pre-existing conditions, the time needed for an EMS unit to arrive at the scene (response time) plays a significant role in the survival rate (*4*). This conclusion is also supported by a study done by Yasmin et al. using the Fatality Analysis Reporting System (FARS) dataset (*6*). To date, a clear relationship has not been fully established between clinically significant improvements in patients' outcomes and reductions in EMS time to definitive care (*4*). However, the consensus is that shorter time to definitive care is associated with improved outcomes for those who need emergency care.

Other than identifying key EMS performance measures, quite a few researchers have made efforts to explore ways to model EMS performance. Do et al. conducted a quantile regression analysis to identify significant factors related to response time. In their analysis, potential impact factors were first divided into patient and system levels (*7*). Meng et al. introduced a mixed logistic regression model to predict the fatality risk in work zones. In their model, the uncertainty involved in accident notification and response times was considered (*8*).

Besides the above regression models, several studies have focused on developing models to identify optimal locations for EMS stations. Maximizing the ambulance coverage and minimizing the en-route time/distance are considered as two primary objectives for EMS station location optimization (*9*). The first objective is referred to as the maximal covering location problem (MCLP), which maximizes the demand that may be served (i.e., within a required time or distance) by properly choosing EMS locations (*9*). The second one is referred to as *p*-median or *p*-center problem (*9*). The *p*-median problem tries to minimize demand-weighted total travel distance whereas the *p*-center problem aims to minimize the maximum distance between demand zones and their nearest ambulance station (*9*).

Given the importance of service uncertainty, such as availability of an ambulance, Daskin et al. proposed a maximal expected covering location problem (MEXCLP) model (*10*). By treating the probability of ambulance availability as a station-specific variable, Hogan and ReVelle modified the MCLP model and introduced a "backup-coverage" term in the objective function (*11*). "Backup-coverage" means that another ambulance may cover the unserved demand points left by the initially assigned ambulance. The modified model aims to maximize both the first and the second coverages to improve EMS robustness. Building on this "backup-coverage" concept, Liu et al. proposed a double standard model, which incorporates two service coverage standards (*12*). Additionally, an adjusted MEXCLP was proposed by Batta et al. to further take into consideration the unavailability probabilities associated with each facility location (*13*).

Several other studies investigated ambulance location problems from a multi-objective perspective (*14*, *15*). For example, Daskin et al. proposed a multi-objective model that can balance between the number of facilities and the extra coverage (*14*). Furthermore, Chanta et al. proposed a bi-objective facility location model for rural EMS stations, with service coverage and service equity as the two objectives (*15*). In their model, service equity represents the fairness of the locations of EMS stations relative to patients. McLay and Mayorga concluded from their research that "longer response travel time results in more equitable patterns of survival: patient lives were saved in rural areas at the expense of losing patient lives in the urban areas" (*16*). The aforementioned EMS location studies mostly are focused on issues related to urban EMS, and few of them address the unique characteristics (e.g., low population and volunteer stations) and needs (e.g., long travel distance) of rural EMS.

Data Collection

For the purpose of this study, EMS incidents, EMS stations, and highway network data were collected. A subset of the National EMS Information System (NEMSIS) data bank consisting of South Dakota EMS incident records was obtained from the Eastern South Dakota EMS Data office. The obtained dataset covers the period between 1/1/2013 and 12/31/2013. Figure 1 shows an example of how the EMS incident records are organized in the NEMSIS dataset and how EMS calls are processed.

Each EMS incident record mainly includes time points (such as "Dispatch"), time intervals (such as "ERTime"), odometers, and incident location. "RespTime" in the NEMSIS dataset is referred to as chute time in many other EMS documents, and is the time period from the notification of the EMS dispatcher to the time the ambulance/responding unit starts moving. Thus, the sum of chute time (or RespTime) and en-route time in this study is equivalent to response time. As this study focuses on 911 calls (e.g., traffic crash, cardiac arrest, chest pain,

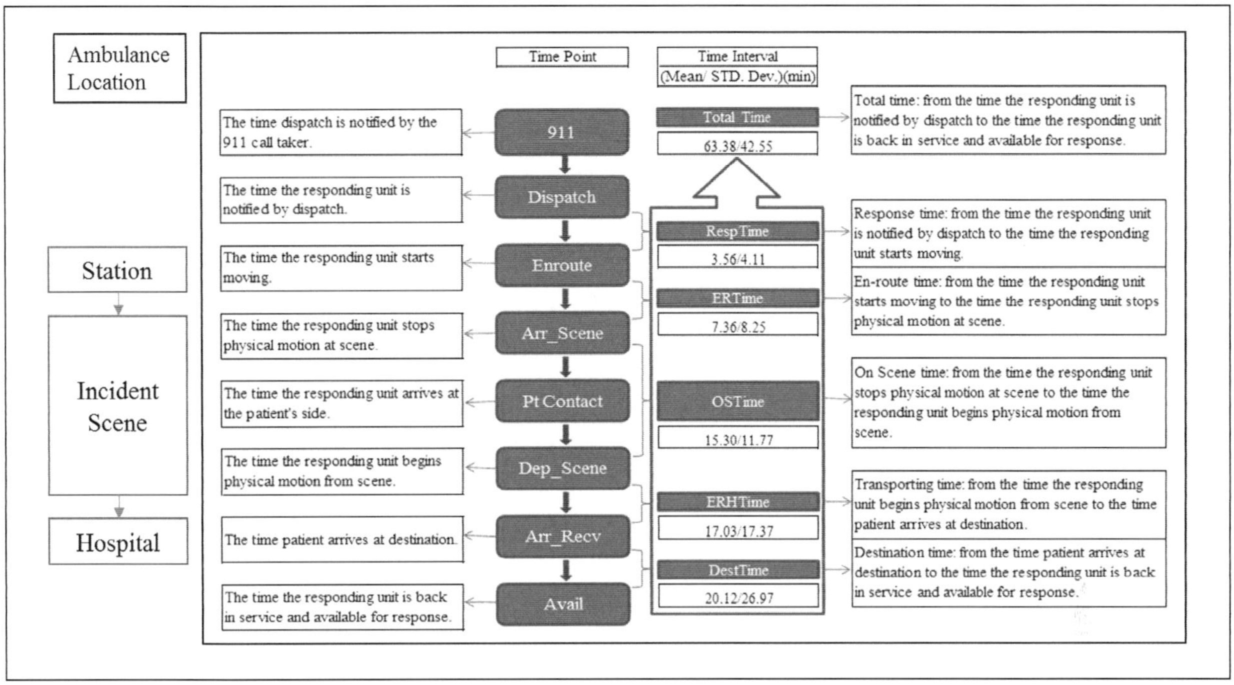

Figure 1. Timeline of EMS process.

animal bite, fall victim, ingestion/poisoning), the original dataset was processed and resulted in 36,198 qualified records in 2013. Google Map API was used to convert the addresses of these 911 incidents into coordinates.

The EMS ambulance station data were obtained from the South Dakota Emergency Medical Services website. This dataset contains information such as name, location, professional status, and vehicle counts for each EMS station. Figure 2 shows the locations for the 36,198 911 incidents and 109 EMS stations in 2013.

The Non-State Trunk Road Inventory (NSTRI) dataset provided by South Dakota Department of Transportation (DOT) was used to derive the highway data. NSTRI includes both interstate highways and local roads in South Dakota. The speed limit information in this dataset was used to perform the network-based analysis. A calculated average en-route speed of 35 mph was used as the travel speed for roads with missing speed limit information.

Spatial Assessment of Statewide EMS Stations

As shown in Figure 2, it seems that 911 incidents cluster around EMS stations. A statistical technique called cross-K function was applied to analyze the 911 incident and EMS location data to confirm the visual assessment. The cross-K function is well suited to analyzing the co-location pattern between two kinds of points. For example, given two sets of locations A $(a_1, a_2, \ldots, a_i)$ and B $(b_1, b_2, \ldots, b_j)$, the cross-K can be used to answer whether the two sets of points are clustered, dispersed, or randomly distributed (*17*). The null hypothesis is that all the points in A are randomly distributed following a binomial point process regardless of the locations in B (*17*). The cross-K function and the corresponding L function are formally defined as

$$K^{ba}(r) = \lambda_a{}^{-1}E$$
$$\text{(number of points } A \text{ within distance } r \text{ of a point in } B) \quad (1)$$

$$L^{ba}(r) = \sqrt{K^{ba}(r)/\pi} \quad (2)$$

where

π is a mathematical constant

λ_a = Density (number per unit area) of points in A;

$E(\ )$ = expected value of A following binomial point process for each point in B;

$K^{ba}(r)$ = K function of A relative to B, for the binomial point process; and

$L^{ba}(r)$ = L function of A relative to B, for the binomial point process.

The expected value $\bar{L}(r)$ $-r$ can be plotted with upper and lower 5% boundaries, which indicate a 90% confidence interval using the Monte Carlo simulation. If $L^{\text{obs}}(r)$ $-r$ is greater than the upper boundary, the co-location pattern can be considered to be significantly

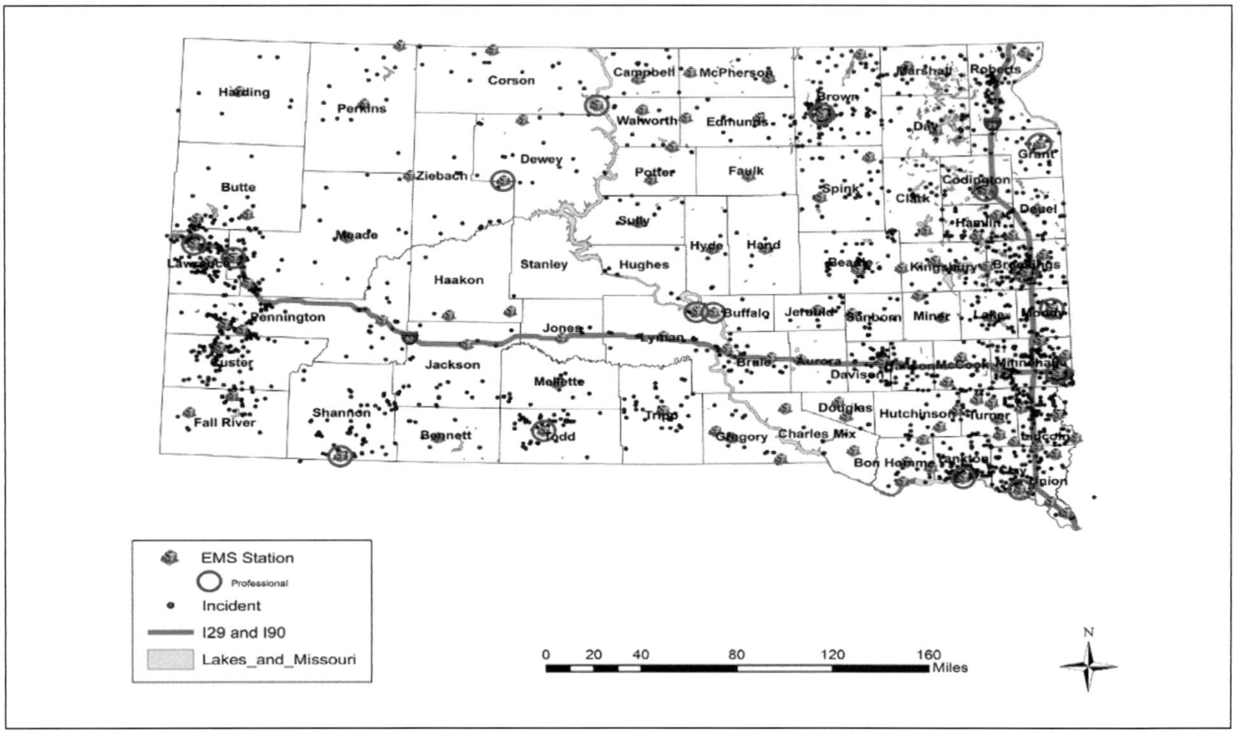

Figure 2. Incidents and EMS stations in South Dakota in 2013.

clustered. If $L^{obs}(r) - r$ is below the lower boundary, the pattern can be interpreted as significantly dispersed. If $L^{obs}(r) - r$ is between the upper and lower boundaries, the points in A and B can be considered to be randomly distributed.

The cross-K function was applied to examine the co-location pattern between 911 incidents (location A) and EMS stations (location B) by using the R software. The results are shown in Figure 3, where $L^{obs}(r) - r$ (represented by the solid black curve) is located above the 5% upper boundary when the distance r is shorter than 25 mi. This suggests that there is a strong co-location pattern (i.e., significant clustering) between EMS stations and 911 incidents if one considers a distance range of 25 mi or less. When the chosen distance is greater than 25 mi, the spatial association between EMS stations and 911 calls becomes statistically insignificant. As most 911 calls are with 25 miles of an EMS station as shown in Figure 2, the current EMS stations overall seem to be well positioned and are in line with where the 911 calls may occur.

Location Optimization of EMS Stations

Facts for Rural EMS

Before formulating the optimization model, the following important facts of rural EMS are analyzed.

Volunteer EMS. Lack of professional staff has long been a major issue affecting EMS service performance in rural areas. Studies indicate that approximately 75% of EMS providers in rural areas are volunteers whereas only 30% are in urban areas (*18*). In South Dakota, less than 20% of the EMS stations have professional personnel (*19*). Not surprisingly, the EMS data show that chute time ("RespTime") was significantly shorter for stations with professional staff than for those staffed by volunteers (2.76 min versus 4.53 min). This longer chute time for volunteers is mainly caused by their need for additional time to go to EMS stations for ambulances and necessary equipment. In the proposed optimization model, the extra time needed for volunteers to go to EMS stations was considered when calculating the chute time.

Low and Sparsely Distributed Service Demand. When optimizing the locations of EMS stations for urban areas, one important consideration is the availability of ambulances because of the high EMS demand. However, for rural EMS, ambulances are available for most of the time. Unit hour utilization (UHU) is defined as the percentage of time an ambulance unit is occupied. It indicates how busy an EMS station is. Based on the 2013 data, most stations in South Dakota had a UHU less than 2%, and the highest UHU was around 10%. This suggests that most EMS stations were not busy and many stations did

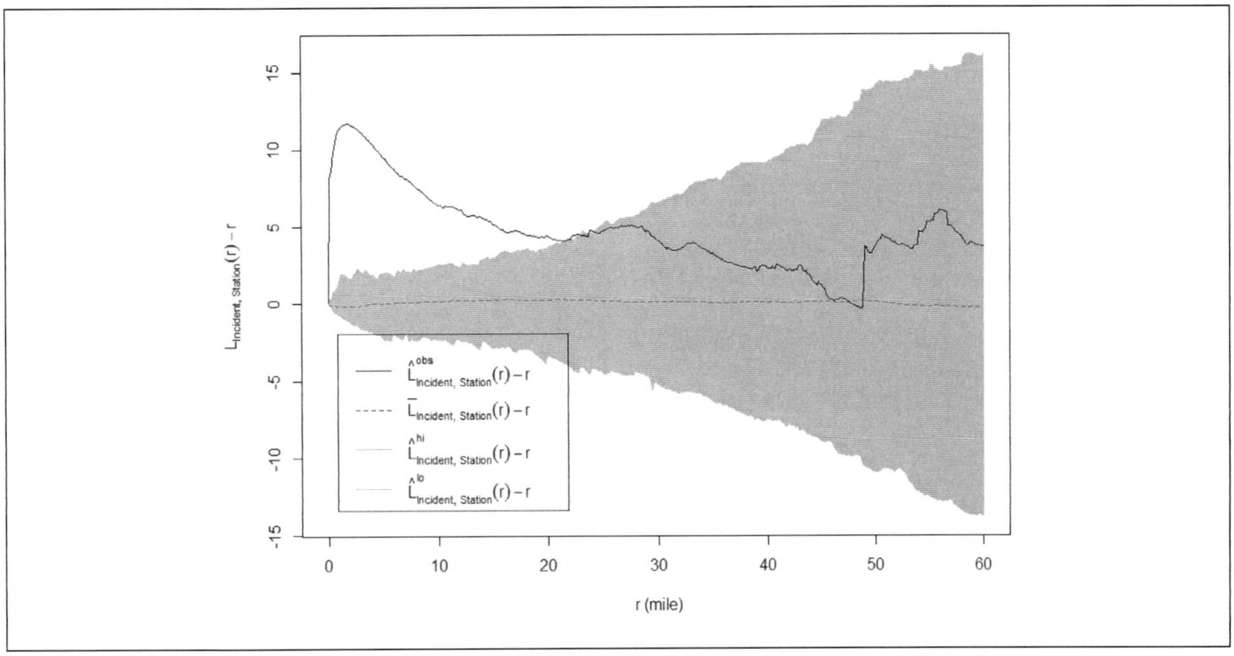

Figure 3. Cross-*K* function and *L* function for 911 incidents and EMS stations.

not have any dispatch request for several days. Even during the busiest days, the service demand was relatively low. Thus, the service/ambulance availability was not considered in the model development in this study.

Long Travel Distance. Besides volunteer EMS, and low and sparsely distributed EMS demand, long travel distance is another major challenge for rural EMS. Unlike populous urban areas, disparities exist in the access to rural EMS stations. In addition, uncovered areas present another challenge as they increase the likelihood that a patient may not be served within the response time standard. This issue is of paramount importance because the patient survival rate is directly related to response time. To account for this factor in the optimization model, two objectives were considered. One is to maximize the size of the covered areas that can provide timely EMS service, and the other is to minimize the response time for remote incident scenes to improve service equity.

Formulation of Optimization Model

To account for both the EMS coverage (Objective 1) and the service equity (Objective 2), a bi-objective location model is shown in Equations 3 and 4, where Z_1 and Z_2 represent the two aforementioned objectives. Specifically, Z_1 maximizes the number of 911 calls that can be covered within a prespecified response time and Z_2 minimizes the average response time for uncovered 911 calls that are outside the prespecified response time zone. x_j is a decision variable, which indicates whether a station should be built at candidate location j. Equations 5 and 6 define y_i, which equals 1 only if demand node i is covered by one or more available EMS facilities. p in Equation 7 denotes the total number of available facilities that can be built. Equations 8 and 9 restrict x_j and y_i to being binary variables. Based on the requirement for South Dakota EMS (*20*), a 15-min threshold value was used for response time to determine whether a demand point is covered or not.

$$\text{Objective}: \text{Max}\, Z_1 = \frac{\sum_{i \in I} D_i y_i}{\sum_{i \in I} D_i} \quad \text{(Objective 1)} \tag{3}$$

$$\text{and Min}\, Z_2 = \frac{\sum_{i \in I} D_i (1 - y_i) \min(t_{ij})}{\sum_{i \in I} D_i (1 - y_i)} \quad \text{(Objective 2)} \tag{4}$$

subject to

$$y_i \leq \sum_{j \in N_i} x_j, \ i \in I \tag{5}$$

$$y_i \geq x_j, \ i \in I, j \in N_i \tag{6}$$

$$\sum_{j \in J} x_j = p \tag{7}$$

$$x_j \in \{0, 1\}, j \in J \tag{8}$$

$$y_i \in \{0, 1\}, i \in I \quad (9)$$

where

i, I are the index and set of demand points,

j, J are the index and set of candidate facility locations,

D_i is the 911 demand at point i,

t_j is the estimated chute time for each candidate facility j,

t'_{ij} is the shortest en-route time from demand point i to facility at point j,

t_{ij} is the response time required for demand i to facility j, $t_{ij} = t_j + t'_{ij}$,

T is the time standard within which coverage is expected ($T = 15$),

$N_i\{j|t_{ij} \leq T\}$ is a set of points j that is within a time of T for point i,

p is the number of facilities to be built,

x_j is a binary variable that equals 1 when a facility is built at point j and 0 otherwise, and

y_i is a binary variable which equals 1 if the node i is covered by one or more facilities and 0 otherwise.

Solution to Optimization Model: Genetic Algorithm

Metaheuristics are one of the most effective methods for solving the proposed EMS location optimization problem. Among the popular metaheuristic solutions (e.g., genetic algorithm, simulated annealing, ant colony optimization, tabu search), the genetic algorithm has been proven to be very effective (*19*). Thus, the genetic algorithm was chosen for this study.

The genetic algorithm adopts natural evolution from Darwin's theory of evolution for the optimization algorithm and has been used in many optimization problems including facility location (*19*). Similar to natural evolution, the essence of this algorithm is to improve the offspring using reproduction mechanisms such as crossover and mutations, resulting in offspring with higher fitness functions (*19*).

The basic procedures for the genetic algorithm are shown in Figure 4 (*21*).

- *Phase 1: Create initial population for the solutions (G set of individuals)*

The G set of initial solutions will be created to activate the process.

- *Phase 2: Evaluate the fitness function of each solution in the population*

The fitness function will be calculated for each solution. This will generate a set of fitness function values.

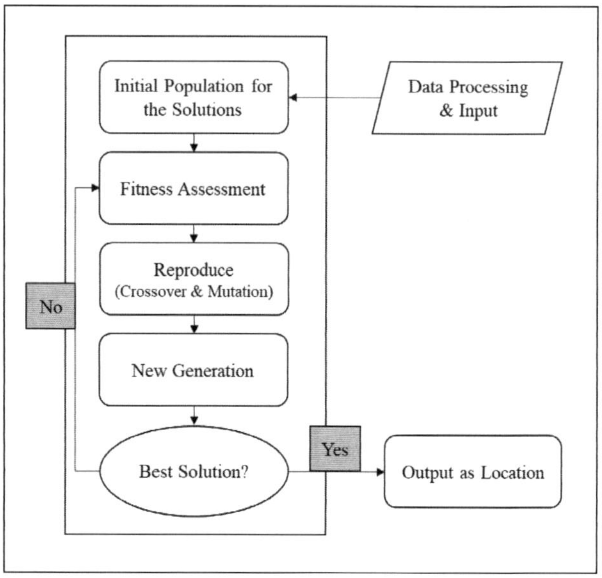

Figure 4. Flowchart for genetic algorithm.

The optimal solution will have the highest fitness function value.

- *Phase 3: Repeat (generate offspring)*

Offspring will be generated in four steps, which are 1) selecting parents from individuals in the population, 2) performing genetic operators (crossover and mutation) to produce new individuals, 3) adding new individuals into the population, and 4) removing individuals with small fitness function values. These steps will be repeated until termination criteria are satisfied.

The genetic algorithm toolbox in the R software was used to solve the proposed multi-objective optimization model. A multicriteria evaluation technique called the weighted sum method was adopted to convert the multi-objective problem into a single-objective one by assigning a weight to each objective (*22*). In practice, the weights can be determined by experts to reflect the relative importance of each objective. In this study, the same weight is used for both objectives, namely $w_1 = w_2 = 0.5$. Equations 10 and 11 show the fitness functions for objectives 1 and 2, respectively.

$$F_1 = Z_1 = \frac{\sum_{i \in I} D_i y_i}{\sum_{i \in I} D_i} \quad (10)$$

$$F_2 = \frac{1}{Z_2} = \frac{\sum_{i \in I} D_i(1 - y_i)}{\sum_{i \in I} D_i(1 - y_i)\min(t_{ij})} \quad (11)$$

These two functions are combined into a single function to normalize the different measurement scales. The combined fitness function is shown in Equation 12.

$F_{1\max}$, $F_{1\min}$, $F_{2\max}$, and $F_{2\min}$ can be obtained by using the genetic algorithm to maximize fitness functions F_1, $\frac{1}{F_1}$, F_2, and $\frac{1}{F_2}$, respectively.

$$F_3 = w_1 * \frac{F_1 - F_{1\min}}{F_{1\max} - F_{1\min}} + w_2 * \frac{F_2 - F_{2\min}}{F_{2\max} - F_{2\min}} \quad (12)$$

Case Study

Data Preparation

Minnehaha County in South Dakota was chosen to demonstrate how to optimize the locations of EMS stations. The data used in this demonstration were obtained using ArcGIS by the following three steps.

1) Create Demand Zone. The "Create Fishnet" spatial tool in ArcGIS was used to create grid cells (demand zones). Here 1-mi by 1-mi cells were chosen to represent the demand zones for 911 calls. The total number of 911 calls inside a cell was counted and assigned to that cell as its attribute (Figure 5).

2) Select Candidate Stations. According to the *National EMS Assessment*, about 40% of EMS agencies are fire departments (*23*). Alternatively, EMS may be stationed at a hospital, a police department, an independent government agency, or a nonprofit/profit corporation. In this study, candidate station locations were selected from 1) the existing stations (Station 1, Station 2, Station 3, and Station 4); 2) hospitals (Station 5, Station 6, and Station 7); 3) a police station (Station 8); and 4) randomly selected sites (Station 9 and Station 10). The randomly selected locations were identified empirically based on the demand zone locations and the road network to increase EMS coverage. This random selection process could be improved by discussing the sites with county EMS officials. The identified candidate stations as well as the demand zones in Minnehaha County are shown in Figure 5. The existing stations are labeled as Station 1 to Station 4 with their volunteer status. Except for the existing stations, it is assumed that all others are volunteer stations.

3) Create Time Matrix. Once the candidate stations were determined, the en-route time matrix was calculated using the network analyst toolbox in ArcGIS. The en-route time matrix is the time needed to travel from the candidate stations to the demand points. It should be noted that chute time differs between volunteer stations and professional stations. For volunteer stations, chute time includes the time required for volunteers to get to a station and the preparation time at the station; for professional stations, on the other hand, chute time is equivalent to the preparation time.

To estimate the average time required for volunteers to get to a station, the authors assumed that the probability of a volunteer staying in one population block equals the percentage of that block's population. The block population data (shown in Figure 6) is obtained from the census website (https://www.census.gov/geo/maps-data/data/tiger-data.html). The estimated travel time for volunteers to get to each station equals the population-weighted average travel time from all population blocks to that station.

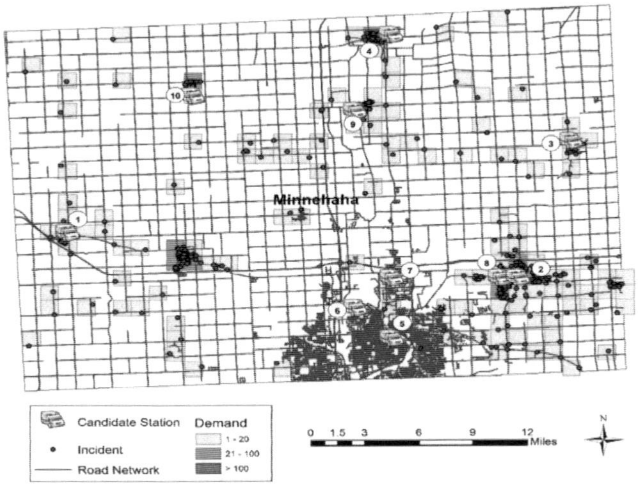

Existing Station	EMS Station	City	Volunteer
1	Humboldt Fire& Ambulance Service	Humboldt	1 (Yes)
2	Med Star Paramedic Ambulance	Brandon	0 (No)
3	Garretson Community Ambulance Corp	Garretson	1 (Yes)
4	Dell Rapids Community Ambulance Service	Dell Rapids	1 (Yes)

Figure 5. Minnehaha County with candidate stations and demand zones.

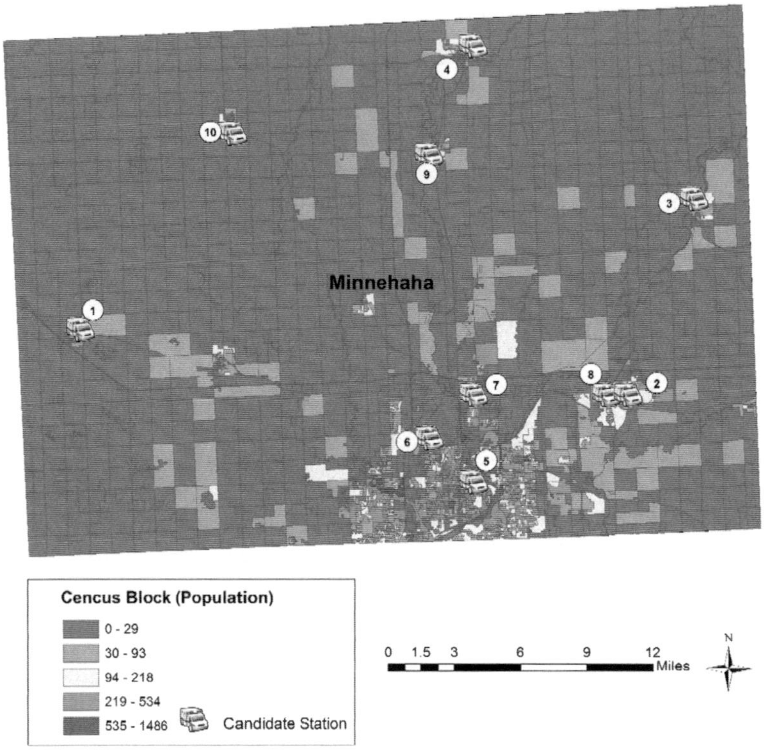

Figure 6. Census block map.

Table 1. Estimated Chute Time for Candidate Stations

Candidate station	1	2	3	4	5	6	7	8	9	10
Volunteer or not (1/0)	1	0	1	1	1	1	1	1	1	1
Estimated time to station (min)	4.17	na	6.25	4.44	3.48	4.01	3.70	5.39	4.61	3.10
Estimated preparation time in station (min)	1	1	1	1	1	1	1	1	1	1
Estimated chute time (min)	5.17	1.00	7.25	5.44	4.48	5.01	4.70	6.39	5.61	4.10

Note: na = not applicable.

Table 1 shows the estimated travel times for each candidate station. The United States National Fire Protection Association (NFPA) 1720 has a 1-min chute time requirement for professional EMS stations. Therefore, 1 min is used as the preparation time for all candidate sites as shown in Table 1 (*24*). The final response time matrix is prepared by adding the chute time to the en-route time.

Model Results and Analysis

Based on the time matrix and the call volume in each demand zone, a genetic algorithm tool in R is used to obtain the optimal solution for siting EMS facilities. To provide a comparison benchmark, a single-objective location model is included, which only considers the EMS coverage (Objective 1). The case study analyzes both single-objective and multi-objective models for a different number of station facilities, which equal 4, 5, 6, and 7, respectively. The results of the optimal locations are shown in Figure 7. As an example, when there are five stations and EMS is made available at candidate Stations 5, 6, 8, 9, and 10, the maximum coverage rate is achieved. When EMS is available at Stations 5, 6, 7, 9, and 10, both service coverage and equity are maximized. By comparing the existing stations (Stations 1, 2, 3, and 4) with the

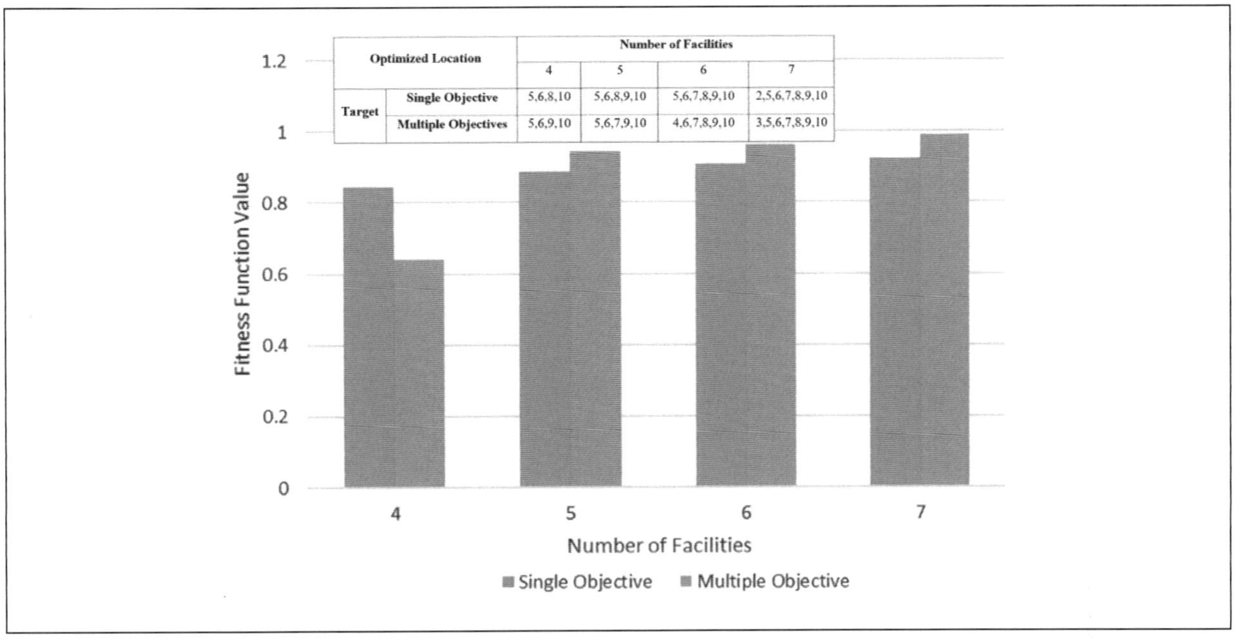

Figure 7. Optimized location EMS stations and the fitness functions.

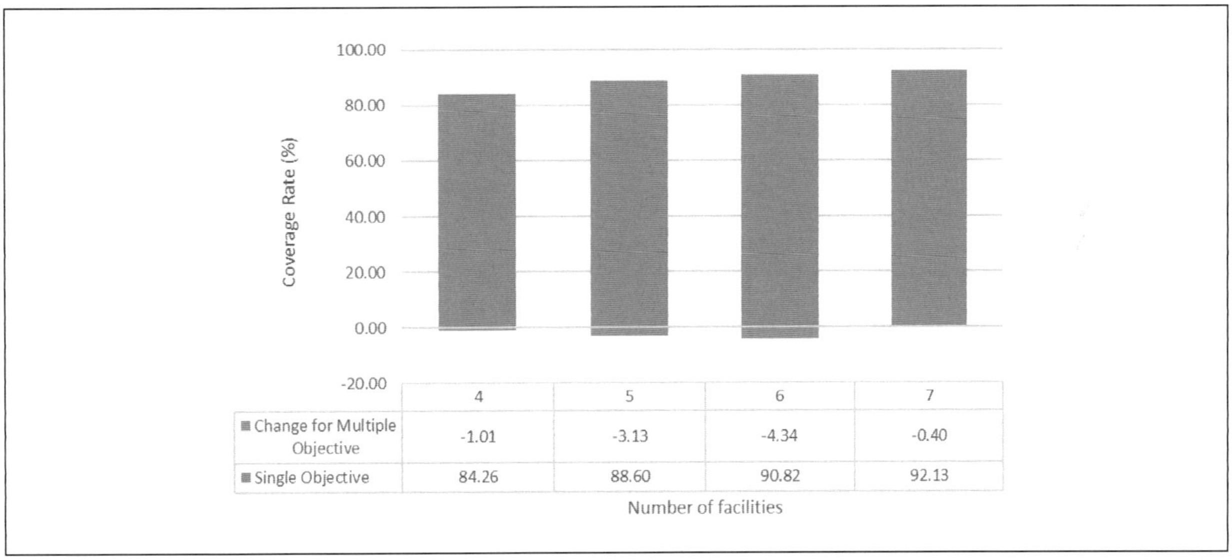

Figure 8. Coverage rate under optimized locations of EMS stations.

optimized stations, significant improvement is observed in both coverage rate and service equity, suggesting the current EMS stations are not optimally located. Figure 7 also displays the trend of the fitness function for different numbers of facilities. It suggests that the fitness functions for both single- and multi-objective models increase as the number of facilities increases. In the single-objective optimization model, service coverage is the EMS service performance measure, which is the fitness function as formulated in Equation 10. In Figure 7, the fitness function value of 0.84 for the single-objective model indicates the coverage rate is 0.84 for the four optimized locations. In the multi-objective optimization model, service coverage and average response time for the uncovered areas are the two service performance measures. The fitness function is now a combination of the two objectives as formulated in Equation 12.

Figure 8 shows the effect of adding service equity as an objective on the service coverage rate for different numbers of facilities. The results from the single-objective

Table 2. Average Response Time (min) under Optimized Locations of EMS Stations

Optimized location		Number of facilities			
		4	5	6	7
Objective(s)	Single objective	14.07	13.55	13.23	12.89
	Multiple objectives	13.88	13.55	14.98	13.19

model are also included here for comparison. After adding the equity objective, an adverse effect on the coverage rate was observed. For the four EMS stations case, the added equity objective decreases the coverage rate by 1%. When the number of facilities increases from four to six, the negative effect becomes more significant with an increase from −1.00% to −4.34%. For the seven facilities case, the negative effect on service coverage, however, is minimal, which may be because there are already enough stations to cover the study area. This observation here suggests that the proposed model is particularly important and useful for rural areas, which usually have fewer EMS stations than urban areas. Before running the model, careful decisions should be made about the respective weights given to service coverage and equity. In this case, each objective was weighted equally. The results could be different if the weights were not the same.

Table 2 shows the average response times calculated based on the optimized location plans. When the number of facilities equals four, the average response time for the multi-objective model is lower than that of the single-objective model.

Conclusions and Future Work

This study sought to propose methods to increase EMS coverage ratio and service equity in rural areas. This was accomplished through the geospatial evaluation of 911 calls and EMS station locations and the optimization of EMS station locations. The spatial associations of 911 calls and EMS stations were confirmed visually as well as using the cross-K function. In the proposed EMS station location optimization model, essential characteristics of rural EMS, such as volunteer stations and long travel distances, were carefully considered. Finally, optimal EMS location solutions were obtained by solving the proposed optimization model using the genetic algorithm toolbox in the R software and the data from South Dakota. The results suggest that the optimization model provides a powerful and practical tool for EMS agencies to strategically plan new stations or relocate existing ones to improve EMS in rural areas.

Although some practical factors related to rural EMS have been carefully modeled in the study, there are still assumptions made that may not accurately reflect real-world rural EMS operations. For example, this study used historical incident location data to represent the 911 demand. In practice, EMS demand may vary by time of day, although this variation is expected to be less significant in rural areas given the overall low traffic demand than in urban areas. To take into consideration the potential EMS demand variation, future work could include the prediction of incident locations and service availability. Furthermore, assumptions have been made to estimate the chute time for volunteer EMS agencies. Future research should thoroughly analyze the chute time for EMS volunteers and simulation models could be used to analyze their dynamics with respect to location, scheduling, and availability.

Author Contributions

The authors confirm contribution to the paper as follows: study conception and design: Qin, He; data collection: He; analysis and interpretation of results: He, Qin; draft manuscript preparation: He, Qin, Xie, Guo. All authors reviewed the results and approved the final version of the manuscript.

References

1. Qin, X., Z. He, and H. Samra. Needs Assessment of Rural Emergency Medical Services. *Transportation Research Record: Journal of the Transportation Research Board*, 2015. 2513: 30–39.
2. American Association of State Highway and Transportation Officials Strategic Highway Safety. *Volume 15: A Guide for Enhancing Rural Emergency Medical Services*, 2005.
3. National Highway Traffic Safety Administration. *Emergency Medical Services Performance Measures: Recommended Measures for System and Service Performance*. National Highway Traffic Safety Administration, Washington, D.C., 2009.
4. Centers for Disease Control and Prevention (CDC). *Motor Vehicle Safety. Injury and Violence Prevention*. http://www.cdc.gov/motorvehiclesafety/index.html. Accessed June 30, 2017.
5. Gitelman, V., K. Auerbach, and E. Doveh. Development of Road Safety Performance Indicators for Trauma Management in Europe. *Accident Analysis & Prevention*, Vol. 60, 2013, pp. 412–423.

6. Yasmin, S., N. Eluru, and A. R. Pinjari. Analyzing the Continuum of Fatal Crashes: A Generalized Ordered Approach. *Analytic Methods in Accident Research*, Vol. 7, 2015, pp. 1–15.
7. Do, Y. K., K. Foo, Y. Y. Ng, and M. E. H. Ong. A Quantile Regression Analysis of Ambulance Response Time. *Prehospital Emergency Care*, Vol. 17, No. 2, 2013, pp. 170–176.
8. Meng, Q., and J. Weng. Uncertainty Analysis of Accident Notification Time and Emergency Medical Service Response Time in Work Zone Traffic Accidents. *Traffic Injury Prevention*, Vol. 14, No. 2, 2013, pp. 150–158.
9. De Smith, M. J., M. F. Goodchild, and P. Longley. *Geospatial Analysis: A Comprehensive Guide to Principles, Techniques and Software Tools*. Troubador Publishing Ltd, 2007.
10. Daskin, M. S. A Maximum Expected Covering Location Model: Formulation, Properties and Heuristic Solution. *Transportation Science*, Vol. 17, No. 1, 1983, pp. 48–70.
11. ReVelle, C., and K. Hogan. A Reliability-Constrained Siting Model with Local Estimates of Busy Fractions. *Environment and Planning B: Planning and Design*, Vol. 15, No. 2, 1988, pp. 143–152.
12. Liu, Y., A. M. Roshandeh, Z. Li, K. Kepaptsoglou, H. Patel, and X. Lu. Heuristic Approach for Optimizing Emergency Medical Services in Road Safety within Large Urban Networks. *Journal of Transportation Engineering*, Vol. 140, No. 9, 2014, p. 04014043.
13. Batta, R., J. M. Dolan, and N. N. Krishnamurthy. The Maximal Expected Covering Location Problem: Revisited. *Transportation Science*, Vol. 23, No. 4, 1989, pp. 277–287
14. Daskin, M. S., K. Hogan, and C. ReVelle. Integration of Multiple, Excess, Backup, and Expected Covering Models. *Environment and Planning B: Planning and Design*, Vol. 15, No. 1, 1988, pp. 15–35.
15. Chanta, S., M. E. Mayorga, and L. A. McLay. Improving Emergency Service in Rural Areas: A Bi-Objective Covering Location Model for EMS Systems. *Annals of Operations Research*, Vol. 221, No. 1, 2014, pp. 133–159.
16. McLay, L. A., and M. E. Mayorga. Evaluating Emergency Medical Service Performance Measures. *Health Care Management Science*, Vol. 13, No. 2, 2010, pp. 124–136.
17. Ripley, B. D. *Spatial Statistics*. John Wiley & Sons, 2005.
18. March, J. A., and J. D. Ferguson. Volunteer EMS Systems. *Medical Oversight of EMS*, Vol. 2, Kendall/Hunt Publishing, 2009, pp. 239–244.
19. Sivanandam, S., and S. Deepa. *Introduction to Genetic Algorithms*. Springer Science & Business Media, 2007.
20. Safetech Solutions LLP. *An Introduction to Rural EMS Medical Direction in South Dakota*. The South Dakota Office of Rural Health, 2011.
21. Sivaraj, R., and T. Ravichandran. A Review of Selection Methods in Genetic Algorithm. *International Journal of Engineering Science and Technology*, Vol. 3, No. 5, 2011, pp. 3792–3797.
22. Grabaum, R., and B. C. Meyer. Multicriteria Optimization of Landscapes Using GIS-Based Functional Assessments. *Landscape and Urban Planning*, Vol. 43, No. 1, 1998, pp. 21–34.
23. Mears, G., B. Armstrong, A. Fernandez, N. Mann, K. McGinnis, C. Mears, N. Sanddal, T. Sanddal, and F. Shofer. *National EMS Assessment*. Report No. DOT HS 811 723. National Highway Traffic Safety Administration, Washington, D.C., 2011.
24. Hensler, B. NFPA 1710, 1720 and Response Times. *FireBureau.com*, 2008.

The Standing Committee on Transportation Safety Management (ANB10) peer-reviewed this paper (18-05425).

Article

TRR
JOURNAL OF THE TRANSPORTATION RESEARCH BOARD

Transportation Research Record
2018, Vol. 2672(32) 94–106

DOI: 10.1177/0361198118798293
journals.sagepub.com/home/trr
SAGE

Land Suitability Analysis for Emergency Medical Services Posts along State Highways: A Case Study of California

Ernest O. A. Tufuor[1], Laurence R. Rilett[1], Yunwoo Nam[1], and Antonio Hurtado Beltran[1]

Abstract

The quicker the response time of emergency medical services (EMS) to road crashes the higher the likelihood of survival. For example, the California Strategic Highway Safety Plan highlights the need to improve the response time and recognizes that 37% and 8% of the fatal crashes are 30 or more miles away from a trauma center in rural and urban areas, respectively. This paper seeks to: (1) demonstrate the viability of using spatial multi-criteria analysis in road safety management; and (2) provide a logical and consistent approach for identifying potential EMS locations. This paper adopted a multi-criteria analysis using the weighted linear combination method on raster data of various impact factors. The area selection criteria were: (1) closeness to probable road fatality locations; (2) distance from existing trauma centers; (3) proximity to existing rest stops; and (4) land availability. An EMS response time of 10 minutes, which is 3 minutes shorter than the national average, is used in this paper for demonstration purposes. Note that the specific criterion will be a function of the decision maker's goal. The proposed methodology, which is a first step in identifying potential areas for EMS locations, was demonstrated on information from the state of California. It was found that of the 155,779 square miles of the land area studied, 24% were suitable, 7% were moderately suitable, and 69% were unsuitable for new EMS locations. In general, the rural areas along the west side freeway corridor, linking Los Angeles and San Francisco, are highly suitable.

The level of injury severity and the probability of the death of people involved in road accidents have been strongly correlated to the emergency response time (*1*). This means that, as the response time decreases, survival rates increase. For example, the probability of surviving a sudden cardiac arrest decreases by about 10% for every minute of delay in treatment (*2*). The response time provides a useful metric for identifying areas for emergency medical services (EMS) locations.

The reports from the Fatality Analysis Reporting System (FARS) of the National Highway Traffic Safety Administration (NHTSA) have shown that EMS response times in fatal road accidents in the United States have improved over the past few years. The significant improvements have occurred in urban settings where the ambulance or hospital accessibility are often close to accident locations. However, this trend has not been found in rural areas in the state of California (*3*). California's Statewide Integrated Traffic Records System (SWITRS) defines a rural area as unincorporated and incorporated areas with a population of fewer than 2,500. Areas with a population of 2,500 or more are classified as urban (*3*).

California is the most populous state in the United States with a population of about 39.1 million (*4*). There were 3,176 fatal crashes reported in 2015, which was the second highest in the United States (*5*). Traffic crashes cost California more than $22 billion every year (*6*). The average response times for EMS were 43 minutes and 5.92 minutes for rural and urban locations, respectively (*7*). In addition, the California Strategic Highway Safety Plan (SHSP, 2015–2019) reports that 37% of fatal collisions in rural California and 8% in urban areas are 30 or more miles away from a trauma center (*3*). There are many ways of providing solutions to bridging this disparity. One possible solution is to locate more EMS services in rural areas with high crash rates under the assumption that this will reduce the EMS response time.

The objective of this paper is to develop a methodology for identifying potential areas for EMS locations. The goal is to achieve a maximum EMS response time of

[1]University of Nebraska-Lincoln, Lincoln, NE

Corresponding Author:
Address correspondence to Ernest O. A. Tufuor: ntc-etufuor@unl.edu

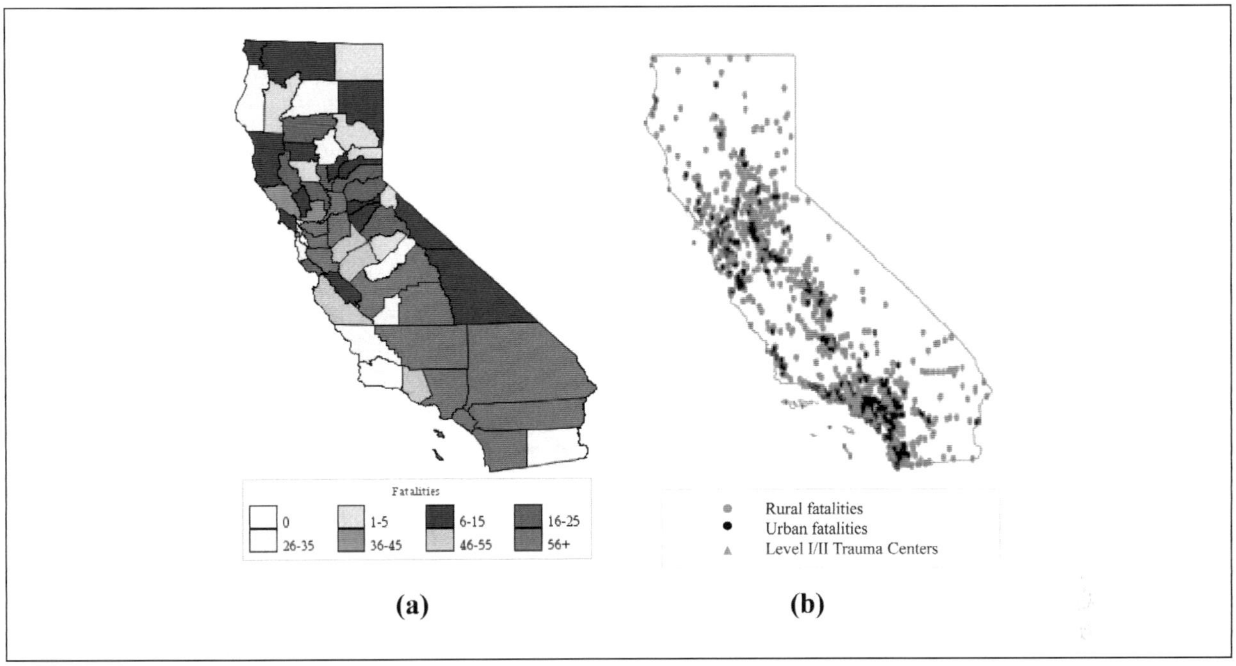

Figure 1. Graphic representation of: (*a*) the 2015 fatalities by county (source: [5]), and (*b*) 2011 fatal crashes and trauma center locations (source: [3]).

10 minutes, which is 3 minutes below the national average. Note that this methodology results in a list of potentially suitable areas and further analysis will be required. This paper makes two main contributions. Firstly, it demonstrates the viability of using Geographical Information System (GIS) Multi-Criteria Analysis in road safety management. Secondly, it provides a logical and consistent approach for identifying potential EMS locations. This can provide necessary background information that local EMS agencies (LEMSAs) can use to select and prioritize potential EMS locations for underserved areas. It should be noted that California is the only state where statutory responsibility for the EMS system rests predominately with the LEMSAs (*8*).

It is assumed that identifying and locating EMS services closer to possible fatality sites will facilitate the early treatment of serious injuries and will ultimately reduce the number of fatalities. This will help to achieve California's EMS-funded grant goals (*9*).

Description of Study Area

The state of California has 58 counties and covers 155,779 square miles. It is the largest state by population and the 3rd largest state by land area (*4*). The cities with highest road fatalities include Los Angeles, San Diego, and San Francisco. Rural crash fatalities make up approximately 46% of the total fatalities in the state (*5*). The locations of road fatalities are mostly in the southern and central regions, as shown in Figure 1*a*.

One of the overall goals of the California State Trauma System, which are incorporated in the State Trauma Plan, is to provide timely access to trauma care (*8*). However, there is wide disparity between locations of rural and urban fatalities and level I/II trauma centers (*3*), as discussed earlier and illustrated in Figure 1*b*.

Literature Review

EMS in California

Most road safety plans specifically address the 4Es of safety: education, enforcement, engineering, and EMS. All of these safety dimensions play a part in reducing crash severity levels and, consequently, fatalities. EMS is involved in two ways: (1) quick response time, and (2) high-quality medical care at the scene and en route to a hospital or trauma center. Other contributing factors include accurately locating the site of the incident, and providing instructions to 911 callers before the arrival of the EMS. The most critical factor for reducing fatality rates is arriving on the scene quickly (*1*).

The SHSP (2015–2019) reports that "one of the reasons fatalities have decreased over the last 20 years is due to the advancements in emergency services." Historically, the focus has been on getting treatment to injured persons at a hospital within the first hour, which is called the "golden hour." However, it is important to remember that the quality of treatment provided by the first responders plays an essential role within this golden hour.

The strategy for EMS as outlined in the SHSP is as follows: (1) increase involvement by EMS leaders in the California SHSP; (2) develop strategies to improve the time to definitive care; (3) improve data from the time of the crash; and (4) improve access to information to enable interoperability of communications, and improve safety at the crash scene. This paper is focused on developing strategies to improve the time to definitive care by providing a reliable and unbiased method in identifying potential areas for new EMS locations.

Multi-Criteria Analysis

Investigating suitable land-use locations within a given area requires a versatile tool that will allow visual and statistical modeling of the interaction among the various critical factors. In this paper, GIS land suitability analysis, also known as multi-criteria analysis (MCA), was chosen to provide this function. GIS-based land suitability analysis is a tool used to enable planners and engineers to analyze the interactions among locations using environment and developmental factors (*10*). Many GIS programs allow users to express simple spatial operations using map algebra to develop land-use suitability models.

The ArcGIS system was chosen for spatial modeling of the problem although it should be noted that any GIS-based software could be used in the proposed methodology. The ArcGIS system provides a user flowchart approach for building the land-use suitability models. Collins et al. (*10*) identified three major approaches to GIS land-use suitability analysis: (1) computer-assisted overlay mapping; (2) multi-criteria evaluation methods; and (3) soft computing or geo-computation methods.

Many researchers have used the MCA method for land suitability analysis. Basnet et al. (*11*) identified suitable land for animal waste applications utilizing a raster GIS. The factors considered were classified and weighted using the analytical hierarchy process (AHP). The suitability level was found to be a function of the weight distribution of the factors.

Rodman et al. (*12*) examined the land suitability for the placement of wind turbines in the nine-county region in the Greater San Francisco Bay Area of California. The authors considered physical factors (e.g., land availability), environmental factors (e.g., vegetation, endangered plant species, and wetlands), and human factors (e.g., urban and recreation locations).

Charabi and Gastli (*13*) incorporated the FLOWA module to assess the land suitability for large photovoltaic farms. The FLOWA module allows for the incorporation of expert opinions on the weightings and criteria, and uses fuzzy theory to model the uncertainty of the experts' choices. The authors used nine factors and found that the land use, land slope, sand/dust risk, grid proximity, and sensitive areas were the key factors in identifying best sites.

Sanchez-Lozano et al. (*14*) also adopted an MCA approach to evaluate solar farm locations in southeastern Spain. The key selection criteria were planning regulations, protected areas, road networks, railway network, waterway network, and mountainous areas. The weightings were influenced by the effect of each factor on achieving the set objectives of the studies.

Frazier (*15*) undertook a land suitability analysis for an inland port to service the ports of Los Angeles and Long Beach in the Southern California region. The author used eight variables, including the presence of slopes, locations of the nearest railway lines, school proximity, and school population. The authors used the same weightings because it was difficult to estimate the most influential variable.

Literature Review Summary

The MCA evaluation method that relies on the weighted summations or linear combinations has been a popular approach for identifying potential site locations. The reason for the popularity is that: (a) it provides a comprehensive approach for analyzing many criteria; and (b) it can be easily implemented within the GIS environment.

The literature review gave a general idea of the important factors to consider in a land-use MCA method. These effects can be grouped into three categories: (i) physical; (ii) environmental or protected areas; and (iii) human factors. The weightings given to each factor depend on their contributing effect in achieving the site-selection goals. These weights are typically identified by decision-makers through statistically based survey techniques. Several techniques have been used including the AHP pairwise comparison method (*16*, *17*), the swing weights technique (*18*, *19*), and the modified-AHP technique (*20*, *21*).

Methodology

As described in the following sections, there are two key steps in the proposed method:

1. data collection and database development, and
2. geoprocessing model.

Data Collection and Database Development

The following key factors were selected for analysis in this paper: location of fatal crashes; existing EMS trauma I/II locations; highways and rest stops locations; lakes;

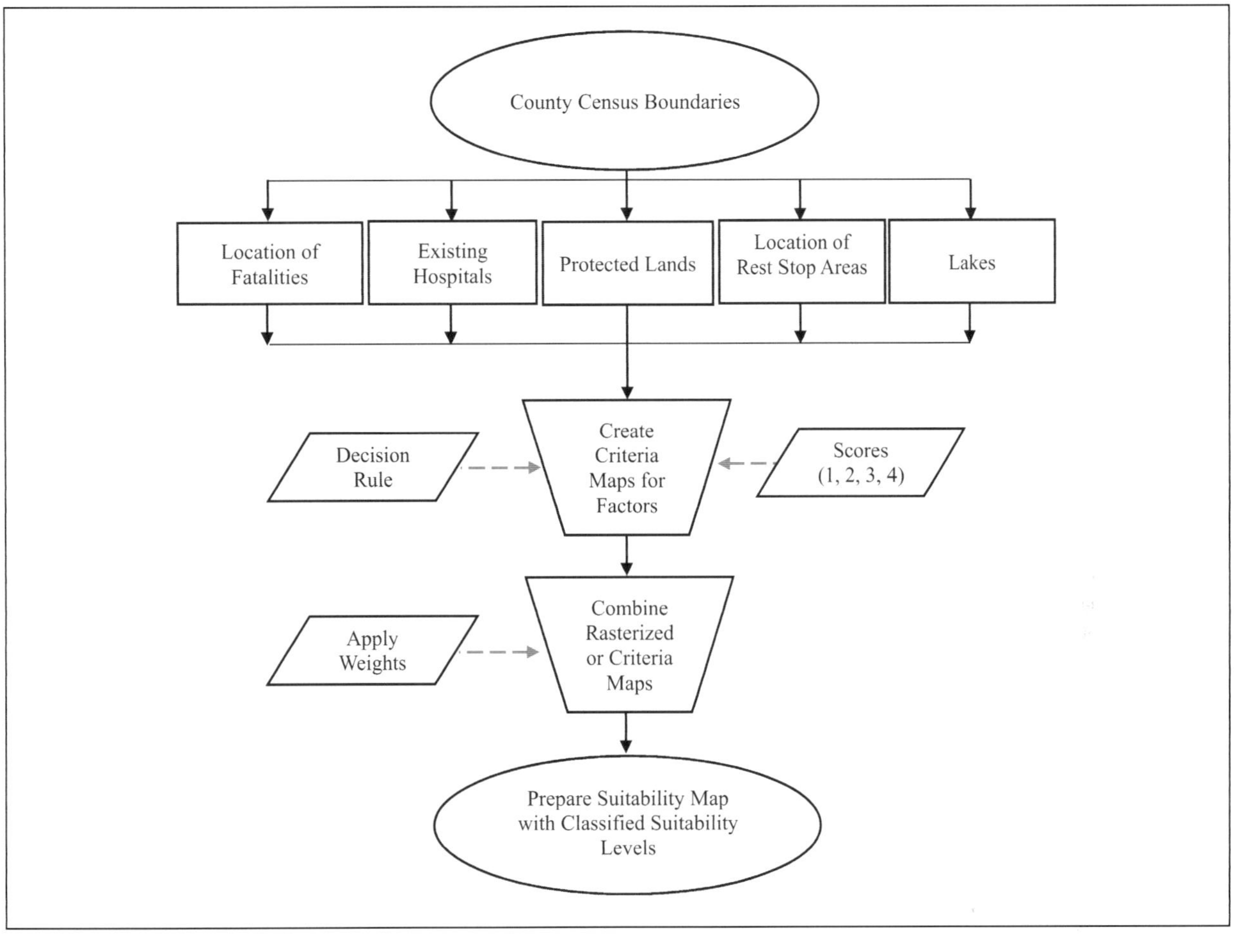

Figure 2. Flow chart for model development.

national parks and reserves; and county boundaries. The sources and a brief description of the data are summarized below.

- The geographical location of fatal crashes was obtained from the University of California, Berkeley's interactive map (*22*). This map allows collision data to be queried and downloaded. A query for 2014–2015 fatalities was undertaken to locate fatal crash locations.
- State highway network and a post-mile GIS databases were used to represent the highways and post-mile point feature classes in California. Each record in the line layer represents a highway segment where the county, route, post-mile prefix, and post-mile suffix are the same, and each segment has a beginning and end post-mile. The post-mile layer displays valid post-mile points at 0.1 (1/10th) mile intervals. Both layers were created based on Caltrans SHN data as of September 4, 2015 (*23*).

Table 1. Decision Rules and Weightings

Decision rules	Influence (%)
Close to fatality points (≤6160 Ft.)	30
Far from existing hospitals	30
Away from protected land areas (≥1000 Ft.)	20
Close to rest stop areas	15
Away from lakes (≥1000 Ft.)	5
Total	100

- Roadside rest areas are already-established suitable land locations for the potential location of EMS services. GIS locations provided by Caltrans (*24*) were used in this paper.
- The geographic location of existing health facilities was obtained from the Office of Statewide Health Planning & Development (*25*). In addition, detailed descriptions of the health facilities were provided. The features used in this study included the hospital type, the nature of the facility (urban

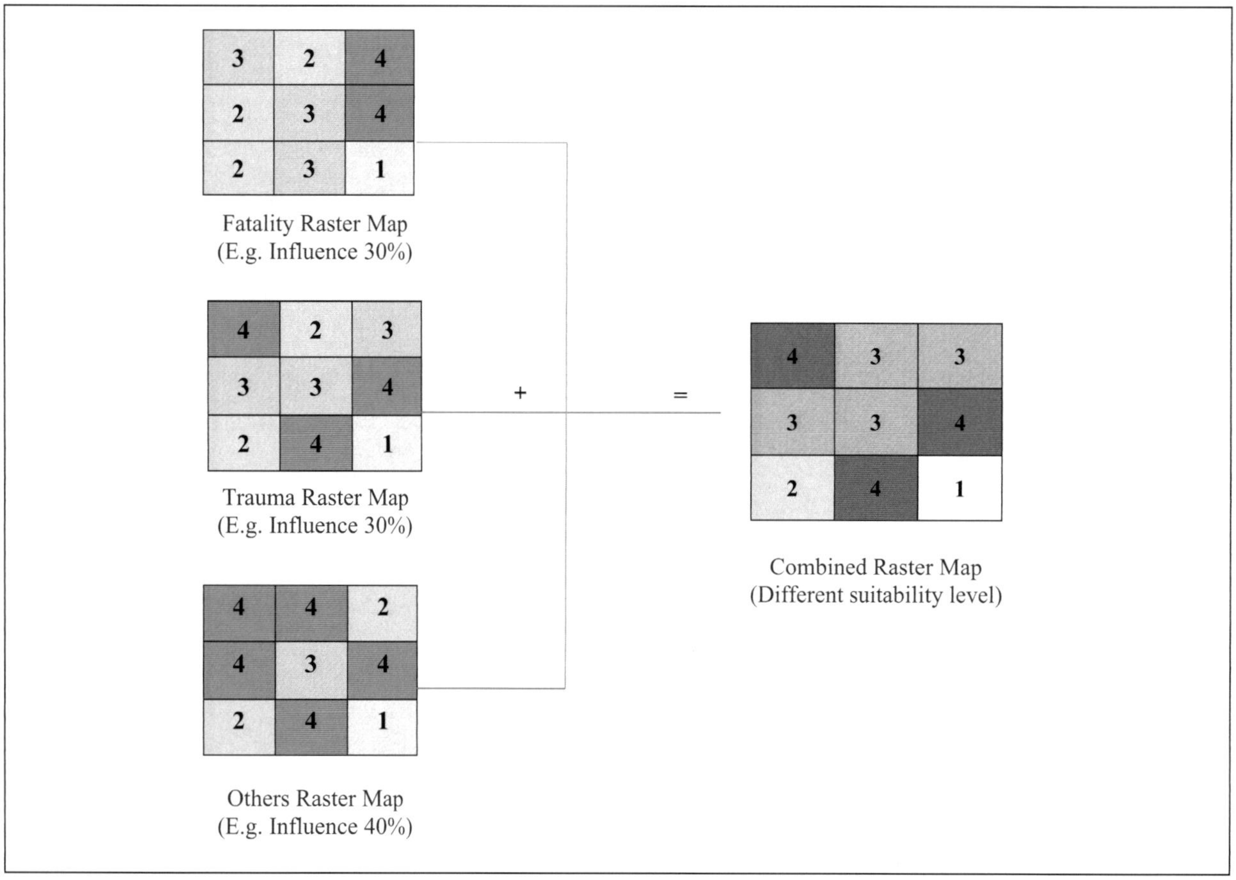

Figure 3. Illustration of the development of the combined raster map.

or rural), and the type of operations (trauma or not).

- The California Department of Fish and Wildlife shapefiles for lakes in California were used to geographically represent the raster map for lake areas. By definition, these areas were avoided during model development. The data is part of the GIS clearinghouse that is reliable and frequently updated (*26*). In addition, a GIS shapefile for the protected areas database (*27*) was also used.

All of the GIS data described above were included as layers using ArcGIS version 10.4. The vector data (e.g., location of fatal crashes, rest stop areas, and locations of existing trauma centers) were mapped and converted to raster format (digitized maps). The conversion from vector to raster formats is discussed in the model development section below. The raster maps were then resampled to a cell size that gave a compactible resolution for the overlay analysis. The process is also discussed in the model development section. The collected data were projected and transformed to the same plane and coordinate system: the North American Datum 1983 State Plane California VI FIPS 0406 (US Feet) and the North American 1983 geographical coordinate system. The purpose of this was to avoid any displacement, offset, or inappropriate overlay of data.

Decision Rules

The maximum speed limit on most California highways is 65 mph (*28*). These speed limits are based on ideal driving conditions. However, in the case of an emergency, where speed in getting to an accident scene or transporting a patient to an emergency medical facility is essential to prevent loss of life, it is plausible to select an upper bound of 70 mph. It should be noted that each LEMSA will need to determine the appropriate ambulance speed level within their specific area for a detailed local base analysis. Considering the EMS target of a 10 minute arrival time at a crash site, this means that areas within a maximum distance of 6,160 Ft. of a crash site will be potential candidates for the location of EMS services. Therefore, the 6,160 Ft. buffer around fatal

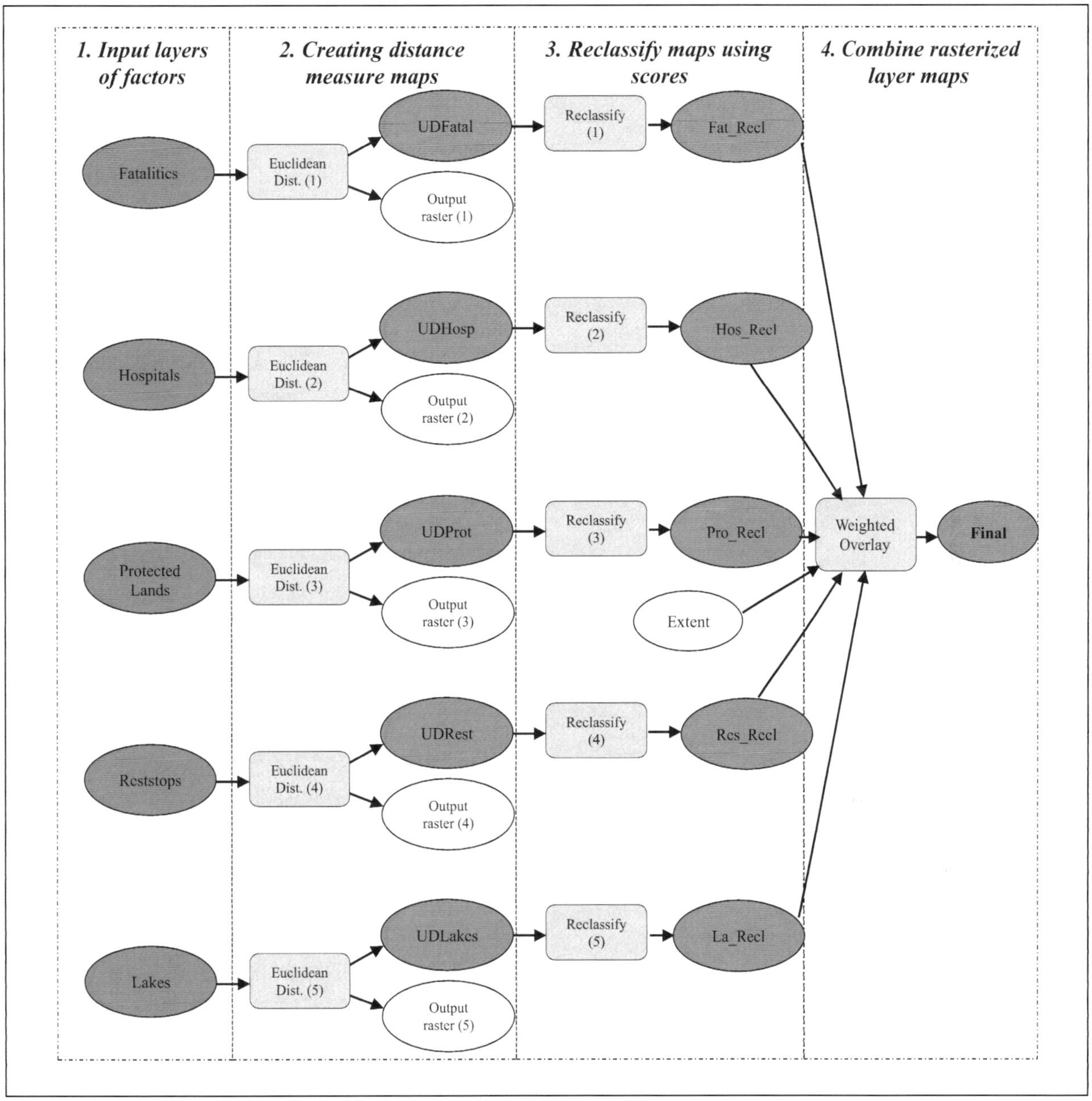

Figure 4. Geoprocessing model.

accident locations is used as the most suitable areas for EMS locations in this paper. Also, areas within 1,000 Ft. of lakes and protected areas were considered unsuitable EMS locations.

Model Development

The model development procedure is illustrated by the flow chart shown in Figure 2. Firstly, the 2010 census data map of every county in California is used as the base map for a spatial representation of each factor (i.e., fatality locations, existing trauma I/II locations, protected lands, rest stop areas, and lakes). A Euclidean distance measure is created on the vector maps to convert them to raster maps. This is done so that the maps will contain the measured distance from every cell to the nearest source or factor (*29*). The cell values on each of the raster maps are reclassified with scores from 1 to 4. The score is used to rank raster cells with regards to the level of relevance. The value of 1 is the least suitable and 4 is the

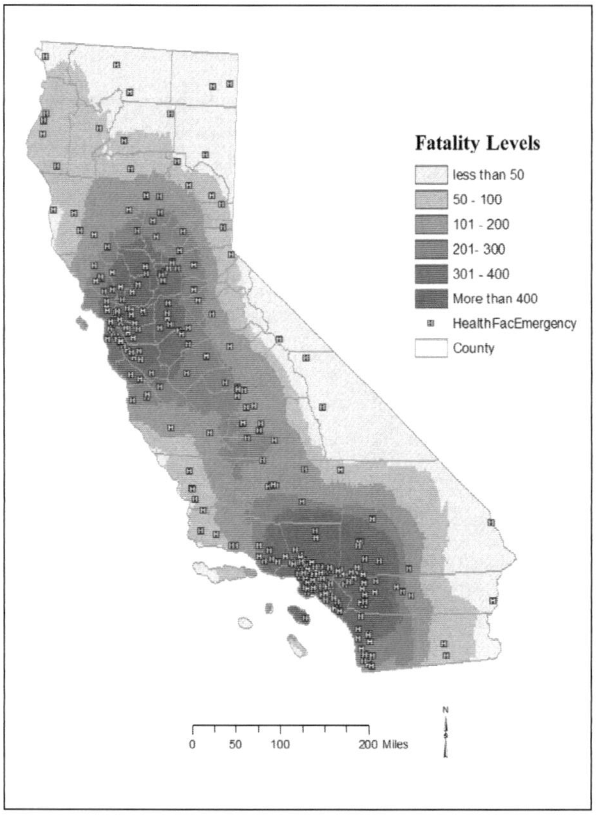

Figure 5. Spatial relationships between aggregated fatal accidents and hospital locations.

most suitable. For example, a cluster of a 6,160 Ft. distance increment is created around each fatality location on the highway. In this case, a score of 1 is used to represent distances greater than multiples of 6,160 Ft. from the rest stop area, and the preferred score of 4 is for areas within 6,160 Ft. of the fatality.

Secondly, the weighted linear combination method is adopted, which is one of the key methods often applied in land-use suitability analysis (*19*). The rasterized layer maps created from each factor are overlaid and combined using predetermined influence percentages or weightings. These weightings are usually selected based on the decision maker's opinion or belief of what the influence of a given factor is in achieving a study objective. For illustration purposes, engineering judgement was used to select the influence levels, as shown in Table 1. For example, areas close to fatality locations (e.g., influence level of 30%) will be preferred over areas close to rest stops (e.g., influence level of 15%). Also, the reason for allocating relatively higher weightings for the protected areas is that buffer zones are often created around those areas as an added layer of protection to the restricted land (*30*). As previously stated, the proposed decision variables, rules, and weightings are examples used to demonstrate the applied methodology. It should be noted that the selection of the type of variables and the influence levels are the responsibilities of decision-makers or the local EMS agency. These are selected based on specific safety objectives. Other possible factors that may be considered include: (1) proximity to access roads; (2) suitable slopes and soil type for construction; and (3) avoiding fire prone locations.

Figure 3 illustrates a typical example of how the combined rasterized layer map is estimated by using only three factors. In the illustration, the raster maps have been reclassified into a common scale using the score of 1–4. The cell values are multiplied by the influence percentage, and the results are added to give the combined raster map. For example, using the upper left cell, the value of the two inputs is $(3 \times 0.3) + (4 \times 0.3) + (4 \times 0.4) = 3.7$. This is rounded to '4' because the weighted overlay uses integer (*29*). Note that this cell is considered a highly suitable cell or area based on its raster score.

Finally, the combined rasterized layer map is "smoothed" using the ArcGIS neighborhood statistics function, which: (1) creates visually attractive maps and (2) helps simplify the analysis. The Environmental Systems Research Institute explains that the function calculates focal statistics for each cell of an image based on a defined focal neighborhood. The developed GIS model layout is shown in Figure 4.

Geospatial Analysis and Results

Figure 5 illustrates a density map of fatal crashes and existing EMS trauma I/II center locations. The ArcGIS density tool distributes the measured quantity of an input point layer throughout the map to produce a continuous surface. The density surfaces show where the aggregated point features (i.e., fatal accident locations from 2014 to 2015) are concentrated. Conceptually, the density function estimates the number of fatalities per unit area from a fatal crash site that falls within the neighborhood around each cell (*29*). It may be seen that most fatal crashes in California are clustered in urban areas (e.g., Los Angeles, San Francisco). This pattern has been documented in the literature (*3*). Not surprisingly, the urban peripherals along state highways linking Los Angeles and San Francisco have a sizable amount of fatal crashes reflecting the high volume of traffic generated by these cities. Similarly, hospitals are also clustered in the urban areas relative to rural areas (*3*). Note that only hospitals that have an EMS unit are considered in this analysis.

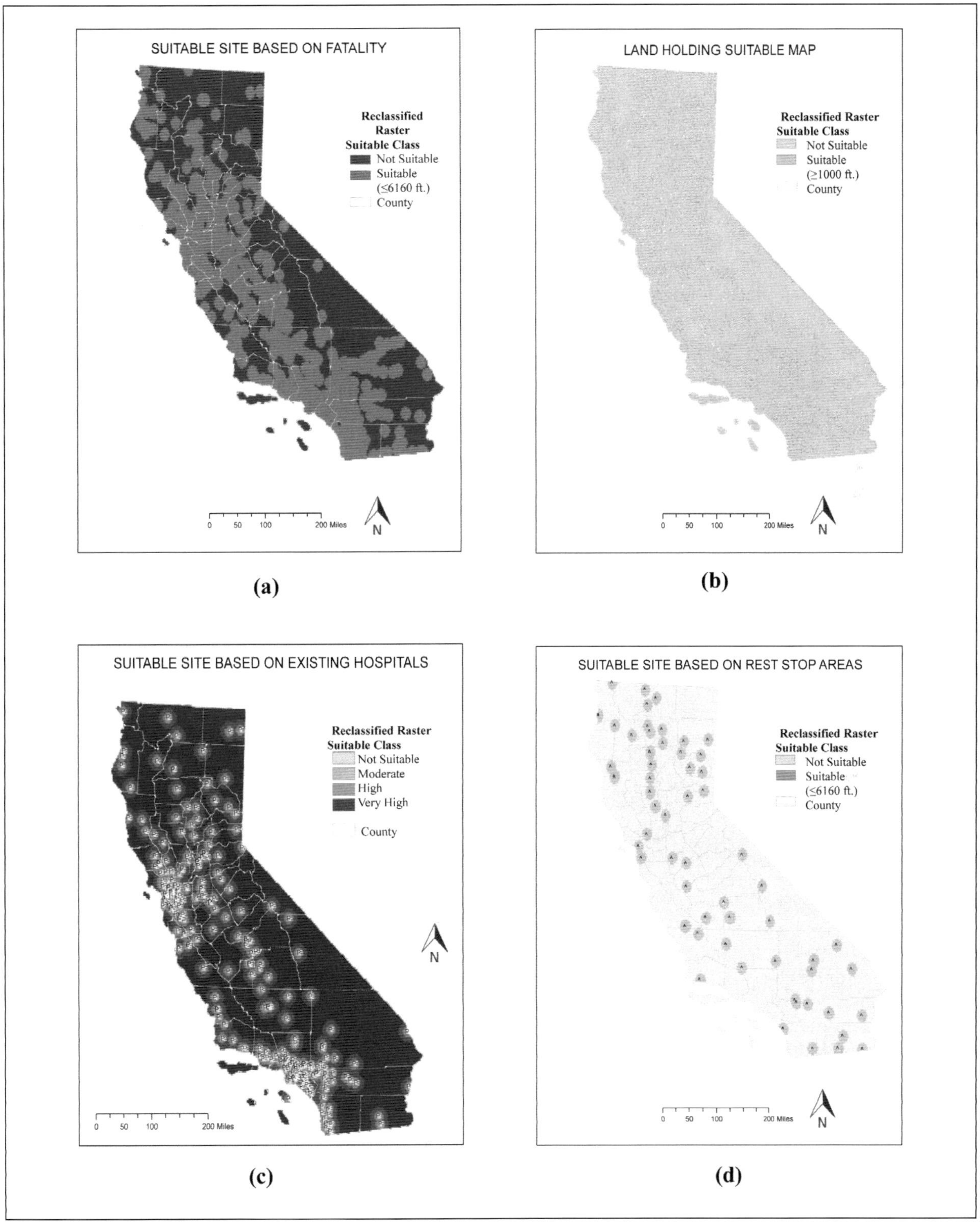

Figure 6. Rasterized maps showing suitable areas per criteria: (*a*) fatalities; (*b*) protected land; (*c*) hospitals; and (*d*) rest stop areas.

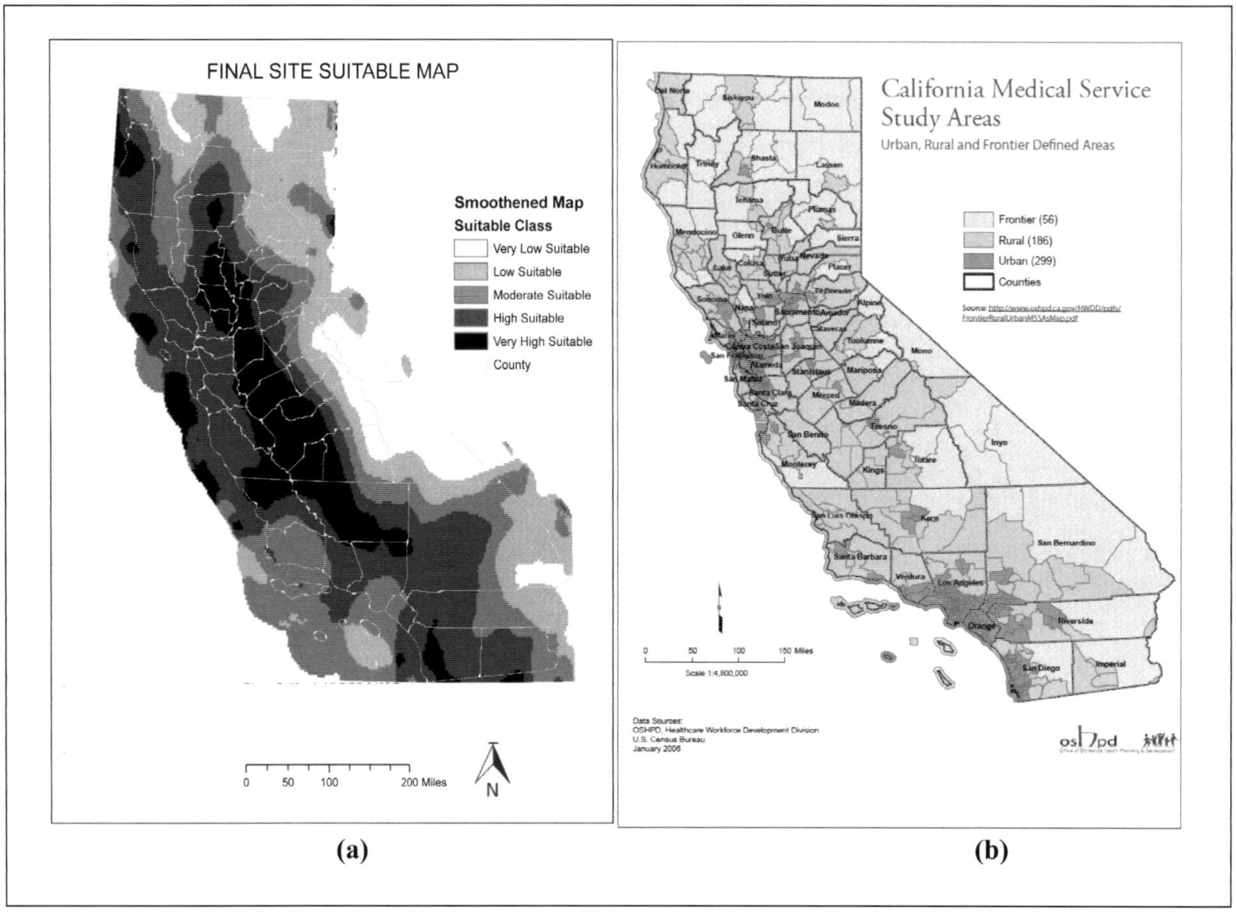

Figure 7. Map showing (*a*) the final combined rasterized EMS post suitability areas; and (*b*) urban and rural areas.
Source (b): Office of Statewide Health Planning and Development.

Raster Maps

Figure 6 illustrates the raster maps for each factor when the decision rules were applied. It may be seen that the suitable areas are generally located in the central and western regions of the state. Not surprisingly, there are some similarities between the land-use map of California (Figure 7*b*) and the raster maps (Figure 6). For example, the land-holding suitability map clearly indicates that California's national parks are unsuitable areas for EMS services, which would be expected given the scoring system described earlier.

Combined Raster Map

In this step, the individual raster maps are combined and smoothed to create a single map that captures the factors and their respective influence levels. From Figure 7*a* it may be seen that the study area is divided into five levels of suitability categories: very low; low; moderate; high; and very high. These categories are based on the total score of each cell in the combined raster map, as previously illustrated in Figure 3.

Figure 7*a* shows the combined raster map when all of the factors and their influence levels are considered. In general, the west side freeway corridor linking the counties between Los Angeles and San Francisco was identified as containing the most suitable areas for rural EMS services. In addition, scattered coastal and southern regions also fall in the high suitability category. It was found that of the 155,779 square miles of land area studied, 37,387 square miles (24%) were classified as suitable, 10,905 square miles (7%) were classified as moderately suitable, and 107,488 square miles (69%) were classified as unsuitable for EMS locations.

Figure 7*b* shows the location of defined urban and rural counties used by the California medical services. This is important to ascertain the rural areas that are within the high to very high suitable category. The details of the identified rural areas are discussed in the remaining sections.

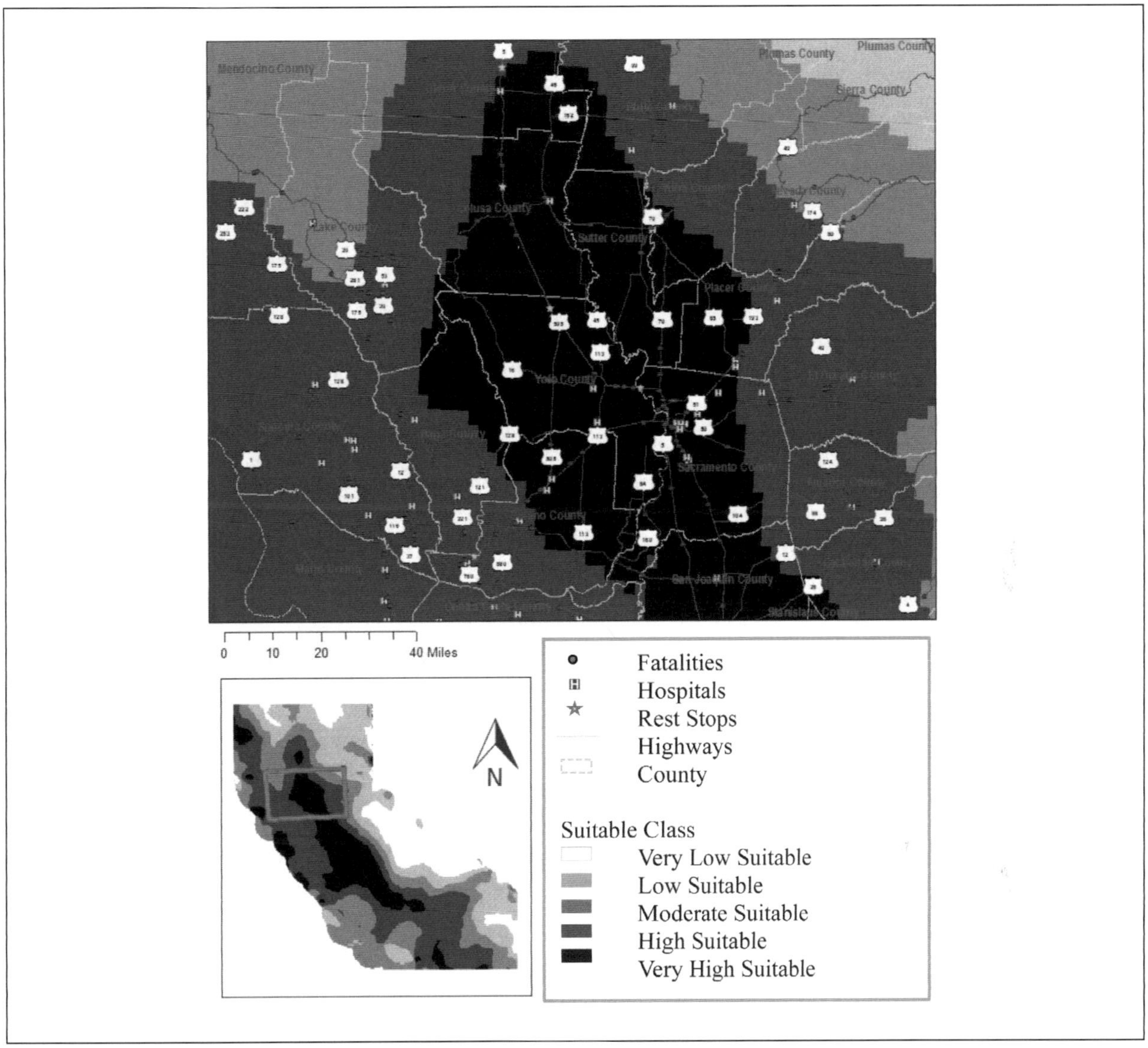

Figure 8. EMS land suitability map along northern California highways.

Figure 8 indicates the suitable land for the northern region. It may be seen that specific locations along Highway 45, Highway 5, and Highway 70 in Colusa County, Sutter County, and Yolo County were classified as suitable rural EMS locations. Predominantly rural counties (i.e., where more than 50% of the county population lives in a rural area) such as Amador, Calaveras (*31*) were all identified as potential rural EMS locations.

Figure 9*a* also shows the suitable land for the central region. These include San Joaquin County, Stanislaus County, and Merced County. From the map, specific locations along Highway 33, Highway 5, and Highway 152 were identified as being suitable.

Very high suitable land for the southern region is shown in Figure 8*b*. Locations along Highways 5, 33, and 41 in Fresno County, Kings County, and Kern County were classified as suitable for EMS locations.

Concluding Remarks

The objective of this study was to develop a logical and consistent approach for identifying potential EMS locations. A response time of 10 minutes, which is 3 minutes shorter than the national average and one-sixth of the "golden hour", is used for demonstration purposes. In actuality, the minimum response time will be chosen by the responsible state agency. In addition,

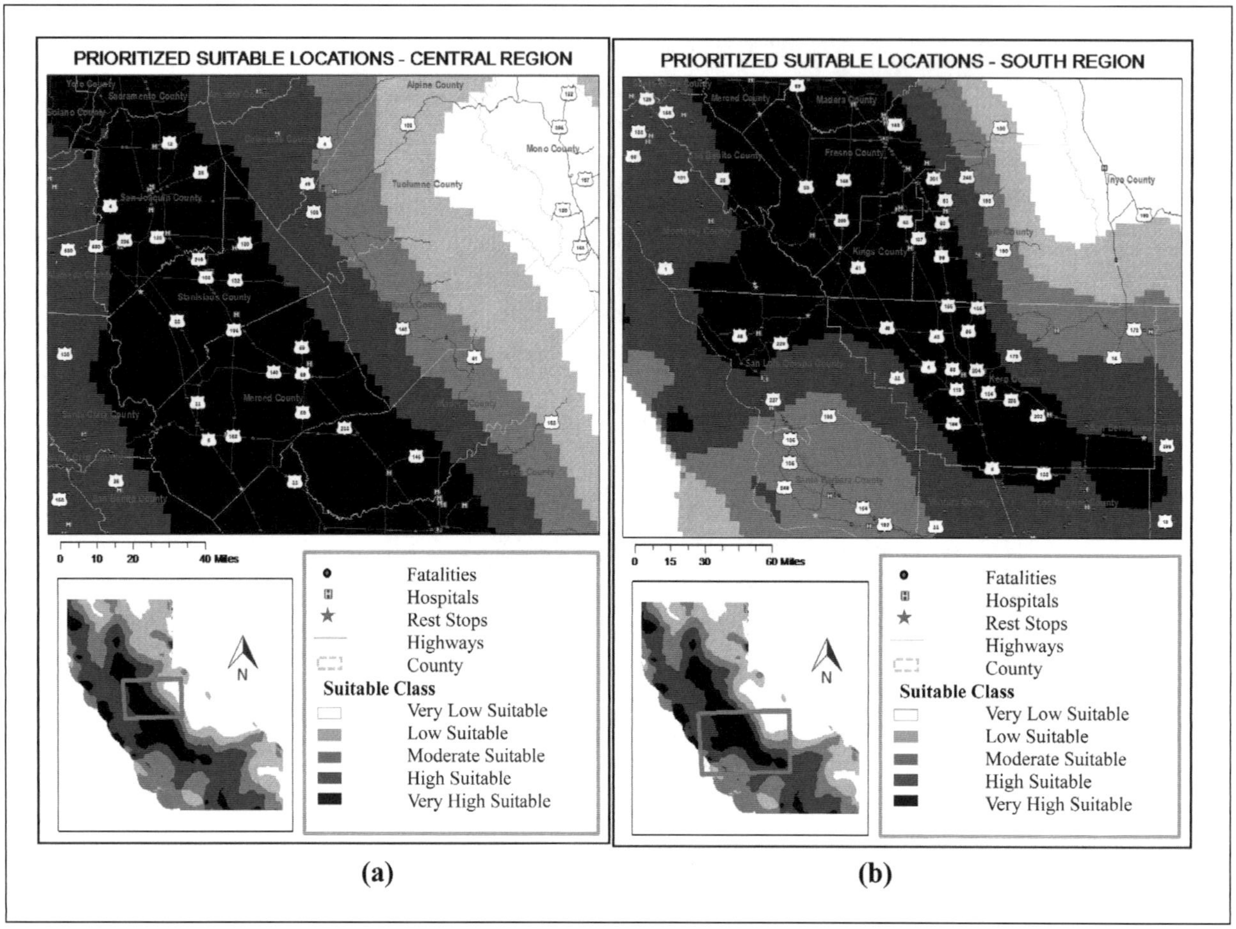

Figure 9. EMS land suitability maps along highways in: (*a*) central; and (*b*) southern California regions.

four criteria (e.g., closeness to probable road fatality locations, distance from existing trauma centers, closeness to existing rest stops, and land availability) and their influence were chosen for illustration purposes. Intuitively, the criteria would be selected by the responsible agencies.

Land-use suitability analysis using GIS has been used in many fields and has been applied in this paper as a tool for road safety management. This paper presented the results of the GIS MCA using the weighted linear combination method. It is hypothesized that the proposed methodology will be useful to LEMSAs because it provides a logical background and justification in selecting viable areas for new EMS locations.

The study found that the west side freeway corridor linking the city of Los Angeles to San Francisco contains the largest size of suitable areas for rural EMS services. A total of 24% of the area is within high to very high suitability levels and about 69% is unsuitable. Preferable locations were along corridors of Highways 5, 33, 41, 45, 70, and 152 within Colusa County, Sutter County, Yolo County, San Joaquin County, Stanislaus County, Merced County, Fresno County, Kings County, and Kern County. Predominantly rural counties such as Amador and Calaveras were also identified as high suitable locations. A cost-benefit analysis is recommended in future studies to determine the suitability of specific sites within the identified counties.

The results provide insight that can be used by decision-makers in identifying potential areas for EMS services. More precise outcomes can be achieved by considering more attributes and improving the weightings pertaining to rural California. Acquiring expert opinion can be used in future research to accurately predict the weightings of the critical factors.

Spatial statistical analysis can be used to estimate the potential fatalities that could be reduced using this methodology. This type of analysis would be useful for the prioritization of EMS post installation at specific locations.

The study was limited to five impact factors. It is recommended that future studies should consider others such as fire prone locations, suitable soil, slopes, and access roads. Hot spot analysis will be essential to determine fatal crash hot spots using three to five consecutive years of fatality data.

Acknowledgments

The authors would like to thank the California State Departments for making GIS data files readily available. The authors appreciate the contributions of Santos Ramos and Amber Hadenfeldt, from the Nebraska Transportation Center, in organizing the manuscript.

Author Contributions

The authors confirm contribution to the paper as follows: study conception and design: Ernest Tufuor, Yunwoo Nam, Laurence Rilett; data collection: Ernest Tufuor, Antonio Hurtado; analysis and interpretation of results: Ernest Tufuor, Laurence Rilett, Yunwoo Nam; draft manuscript preparation: Ernest Tufuor, Laurence Rilett, and Antonio Hurtado. All authors reviewed the results and approved the final version of the manuscript.

References

1. Raj, R., J. Siironen, R. Kivisaari, M. Kuisma, T. Brinck, and J. Lappalainen., and M. B. Factors Correlating with Delayed Trauma Center Admission Following Traumatic Brain Injury. *Scandinavian Journal of Trauma, Resuscitation and Emergency Medicine*, Vol. 21, No. 1, 2013, p. 67.
2. Mississippi State Department of Health. Report to House and Senate Public Health and Welfare Committees. http://msdh.ms.gov/msdhsite/index.cfm/47,2622,313,386,pdf/ambulancestudy.pdf. Accessed February 26, 2017.
3. *SHSP, 2015-2019: California Strategic Highway Safety Plan*. http://www.dot.ca.gov/trafficops/shsp/docs/SHSP15_Update.pdf. Accessed December 7, 2016.
4. US Census Bureau. *California: 2010 Census of Population and Housing*. CPH-2-6. August 2012. https://www.census.gov/prod/cen2010/cph-2-6.pdf. Accessed May 5, 2017.
5. National Highway Traffic Safety Administration. *Fatality Analysis Reporting System*. California Traffic Safety Facts. https://cdan.nhtsa.gov/SASStoredProcess/guest#TAB1. Accessed May 5, 2017.
6. *California Department of Transportation. SHSP Fact Sheet*. Caltrans, 2015. http://www.dot.ca.gov/trafficops/shsp/docs/shsp_fact_sheet.pdf. Accessed November 25, 2016.
7. National Highway Traffic Safety Administration. *Traffic Safety Facts 2014*. DOT HS 812 032. Washington, DC: National Highway Traffic Safety Administration, 2014. https://crashstats.nhtsa.dot.gov/Api/Public/ViewPublication/812261 Accessed December 5, 2016.
8. Emergency Medical Services Authority. *2014 California State Trauma Plan*. #13-301. http://www.emsa.ca.gov/Media/Default/Commission%20Meeting%20Docs/March%202015/6C_State_Trauma_Plan_Attachment.pdf Accessed February 26, 2017.
9. California Office of Traffic Safety. *Highway Safety Plan - 2016*. https://www.nhtsa.gov/sites/nhtsa.dot.gov/files/ca_fy16hsp.pdf Accessed February 26, 2017.
10. Collins, M. G., F. R. Steiner, and M. J. Rushman. Land-use Suitability Analysis in The United States: Historical Development and Promising Technological Achievements. *Environmental Management*, Vol. 28, No. 5, 2001, pp. 611–621.
11. Basnet, B. B., A. A. Apan, and S. R. Raine. Selecting Suitable Sites for Animal Waste Application Using a Raster GIS. *Environmental Management*, Vol. 28, No. 4, 2001, pp. 519–531.
12. Rodman, L. C., and R. K. Meentemeyer. A Geographic Analysis of Wind Turbine Placement in Northern California. *Energy Policy*, Vol. 34, No. 15, 2006, pp. 2137–2149.
13. Charabi, Y., and A. Gastli. PV Site Suitability Analysis Using GIS-based Spatial Fuzzy Multi-criteria Evaluation. *Renewable Energy*, Vol. 36, No. 9, 2011, pp. 2554–2561.
14. Sánchez-Lozano, J. M., J. Teruel-Solano, P. L. Soto-Elvira, and M. S. García-Cascales. Geographical Information Systems (GIS) and Multi-Criteria Decision Making (MCDM) Methods for the Evaluation of Solar Farms Locations: Case Study in South-Eastern Spain. *Renewable and Sustainable Energy Reviews*, Vol. 24, 2013, pp. 544–556.
15. Frazier, E. M. *A Site Suitability Analysis for an Inland Port to Service the Ports of Los Angeles and Long Beach*. Doctoral dissertation. University of Southern California, Los Angeles, CA, 2014.
16. Banai, R. Fuzziness in Geographical Information Systems: Contributions from the Analytic Hierarchy Process. *International Journal of Geographical Information Science*, Vol. 7, No. 4, 1993, pp. 315–329.
17. Saaty, T. L. Decision Making with the Analytic Hierarchy Process. *International Journal of Services Sciences*, Vol. 1, No. 1, 2008, pp. 83–98.
18. Bodily, S. E. *Modern Decision Making: A Guide to Modelling with Decision Support Systems*. New York: McGraw-Hill, 1985.
19. Malczewski, J. On the Use of Weighted Linear Combination Method in GIS: Common and Best Practice Approaches. *Transactions in GIS*, Vol. 4, No. 1, 2000, pp. 5–22.
20. Kritikos, T. R. H., and T. R. H. Davies. GIS-based Multi-Criteria Decision Analysis for Landslide Susceptibility Mapping at Northern Evia, Greece [GIS-basierte multikriterielle Entscheidungsanalysen zur Kartierung von Massenverlagerungspotenzialen im nördlichen Evia, Griechenland.]. *Zeitschrift der Deutschen Gesellschaft für Geowissenschaften*, Vol. 162, No. 4, 2011, pp. 421–434.
21. Nefeslioglu, H. A., E. A. Sezer, C. Gokceoglu, and Z. Ayas. A Modified Analytical Hierarchy Process (M-AHP) Approach for Decision Support Systems in Natural Hazard Assessments. *Computers & Geosciences*, Vol. 59, 2013, pp. 1–8.
22. Transportation Injury Mapping System. *Transportation Injury Mapping System. University of California, Berkeley, CA*, 2016. https://tims.berkeley.edu/tools/gismap_old/#. Accessed November 25, 2016.
23. California Department of Transportation. *Caltrans GIS Data Library*. http://www.dot.ca.gov/hq/tsip/gis/datalibrary/. Accessed November 25, 2016.

24. California Department of Transportation. *California Highway Safety Roadside Rest Areas.* http://www.dot.ca.gov/hq/tsip/gis/datalibrary/Metadata/RestArea.html Accessed November 25, 2016.
25. Office of Statewide Health Planning & Development. *California Healthcare Facilities Downloadable GIS Data Sets.* http://www.oshpd.ca.gov/General_Info/Healthcare_Atlas.html Accessed November 25, 2016.
26. *California Department of Fish and Wildlife.* https://www.wildlife.ca.gov/Data/GIS/Clearinghouse. Accessed November 25, 2016.
27. California National Park and Reserves. *California Protected Areas Database.* http://www.calands.org/data. Accessed November 25, 2016.
28. California Department of Motor Vehicles. *Driver Handbook – Laws and Rules of the Road.* https://www.dmv.ca.gov/portal/dmv/detail/pubs/hdbk/speed_limits+. Accessed December 7, 2016.
29. *Esri ArcGIS Desktop.* http://desktop.arcgis.com/en/arcmap/10.3/tools/spatial-analyst-toolbox/how-point-density-works.htm Accessed May 5, 2017.
30. Martino, D. Buffer Zones Around Protected Areas: A Brief Literature Review. *Electronic Green Journal,* Vol. 1, No. 15, 2001.
31. Stanford Medicine. *eCampus Rural Health.* http://ruralhealth.stanford.edu/health-pros/factsheets/Accessed February 26, 2017.

The Standing Committee on Transportation Safety Management (ANB10) peer-reviewed this paper (18-05729.)

The contents of this paper reflect the views of the authors, who are responsible for the facts and the accuracy of the information presented here, and are not necessarily representative of the state of California, LEMSAs, and any local agency.

Article

TRR
JOURNAL OF THE TRANSPORTATION RESEARCH BOARD

Transportation Research Record
2018, Vol. 2672(32) 107–119

DOI: 10.1177/0361198118799038
journals.sagepub.com/home/trr
SAGE

Impacts of State-Specific Policy and Legislation on Safety Advancement by Departments of Transportation

Jennifer H. Ogle[1], Sababa Islam[1], Kweku Brown[2], William J. Davis[2], and Wayne A. Sarasua[1]

Abstract
The overall goal of this research was to identify proven successful safety programs used in other states and assess the potential for safety improvement if similar programs were implemented in South Carolina. The research team not only sought out engineering solutions, but also expanded the search to include programs for enforcement, education, licensing, legal proceedings, and emergency services—therefore incorporating a wide range of stakeholder groups. South Carolina has, for many years, had one of the highest mileage death rates of any state in the nation—far exceeding the national fatality rate. While South Carolina Department of Transportation has a federal requirement to develop and maintain the Strategic Highway Safety Plan, which identifies the state's key safety needs and guides investment decisions toward strategies and countermeasures with the most potential to save lives and prevent injuries, South Carolina legislation and state policies have effectively blocked many paths to safety improvements. Tree protection ordinances, limited policies for graduated drivers licensing, bans on camera enforcement, and lack of universal helmet laws continue to undermine efforts to improve motor vehicle safety in the state. Using a data-driven approach to safety program selection will yield support for changes in programs, policies, and standards, and have positive impacts on safety, operational, and economic aspects of the South Carolina roadway system. Further, the implementation of a data-driven safety management program will help to assure that the most appropriate strategies are implemented. The successful implementation of this research would likely result in a substantial reduction in loss of life and injuries associated with motor vehicle crashes in the state of South Carolina.

The overall goal of this research was to identify proven safety programs which had been shown to reduce crashes in other states, and assess the potential for safety improvement if similar programs were implemented in South Carolina. The search for proven safety programs began with an evaluation of existing crash problems, population characteristics, and driving environments to determine which areas were most problematic. Initially, the research team not only sought out engineering solutions, but also expanded the search to include programs for enforcement, education, licensing, prosecuting traffic violations, and emergency services—therefore incorporating a wide range of stakeholder groups. The latest Federal Aid Highway Bills, beginning with the Moving Ahead for Progress in the 21st Century (MAP-21) Act (2012) and continuing in the Fixing America's Surface Transportation (FAST) Act (2015), emphasize performance-based measures to qualify for funding in most transportation programs, and much attention was given to prioritizing safety measures throughout the bills. With that in mind, South Carolina Department of Transportation (SCDOT) sought to implement highway safety improvements based on factual data with indications of positive expectations for the improvements.

Although the SCDOT has a large role in safety of the transportation system, it only represents one stakeholder in the safety movement. Other stakeholders include enforcement, education, emergency medical services, licensing, public safety, court systems, and legislative/policy branches of government. To achieve significant gains in the safety of our transportation system, SCDOT and other potential stakeholders must seek collective efficacy to successfully drive down crashes, injuries, and

[1]Clemson University, Clemson, SC
[2]Civil and Environmental Engineering, The Citadel, Charleston, SC

Corresponding Author:
Address correspondence to Sababa Islam: sislam@clemson.edu

fatalities. Progress has been made toward addressing many of our state's safety challenges, thanks in part to research, education, and infrastructure actions, but there remains much work to be done.

In the course of this research, the authors conducted a national search for proven successful safety programs that have resulted in significant reductions in fatal and serious injury collisions in each of the problem areas identified in this research (e.g., serious crash types, high-risk drivers, special vehicles, vulnerable roadway users, etc.). The research team reviewed these programs and selected a portion of the programs for in-depth review and analysis. During the reviews, researchers sought information on the effectiveness of the safety programs in reducing fatal and serious injury collisions, as well as costs and other programmatic requirements. The potential for safety improvement in South Carolina was assessed on several factors including: (a) the magnitude of the problem in the state, (b) geographic boundaries of problem areas, (c) demographic trends, (d) current legislative statutes, (e) state regulations and agency policies, and (f) other factors that contributed to the success of the program.

In general, the assessed countermeasures fell into two categories—behavioral and infrastructure related. While comparing infrastructure safety countermeasures that other states have been implementing to those used in South Carolina, researchers found that SCDOT was effectively utilizing most of the proven infrastructure treatments. Examples of successful infrastructure countermeasures include: rumble strips, median cable barriers, oversized signs, and high-friction surface treatments. However, one area where SCDOT has had significant crash history and limited safety improvement is in regard to roadside trees. In many states, authority is given outright to the DOT to maintain clear zone areas within the right-of-way, but numerous tree ordinances put into place by cities and counties in South Carolina require extensive processes to request permission to remove trees. The ultimate decision to remove a tree from the clear zone is often out of control of SCDOT. In this case, there is a conflict between policies meant to protect the natural beauty and resources in the state and those meant to protect motorists who inadvertently find themselves leaving the safety of the travel lanes. On the behavioral side, many successful programs in other states were either minimally adopted by South Carolina, or altogether lacking. Although South Carolina does have graduated drivers licensing, driving under the influence (DUI), and partial helmet use laws in place, many of the component parts of the legislation are quite lenient, and other programs such as automated speed enforcement have been effectively banned by legislation.

When researchers looked at these programs in totality, it was found that, in many cases, practices, policies, and legislation in South Carolina did not support the best practices in safety programming. An initial assumption of the team was that some of the programs that would be identified as highly successful in other states may have limited impact in South Carolina because of current legislation and policies. In hindsight, South Carolina legislation and policies turned out to be the top factor impeding the introduction of proven safety programs. This research reviews some of the safety problems in the state, the current legislative and policy stance, and required changes in policies to mimic other states with successful safety programming initiatives. Finally, the research provides a return on investment (ROI) to show the potential impacts of legislative and policy decisions within the state.

Background

For many years, South Carolina has had one of the highest mileage death rates of any state in the nation and far exceeding the national fatality rate. On average in 2015, almost three people were killed in a crash every day in South Carolina, and a traffic crash was reported roughly every four minutes. Traffic collisions are responsible for billions of dollars in economic loss to South Carolina each year. Economic loss includes property damage, medical costs, and lost productivity, but does not include intangible costs such as grief and suffering. In the year 2015, fatal traffic crashes in South Carolina result in over $4 billion in economic loss (*1*).

In 2015, there were 911 fatal collisions, 37,861 injury collisions, and 95,189 property damage-only (PDO) collisions in South Carolina. Traffic fatalities increased 19% between 2014 and 2015. There were 3.684 million licensed drivers, 4.163 million registered vehicles, and a total population of 4.896 million in South Carolina in the same period (*2*). Following is the list of the 2015 national fatality rate rankings that shows South Carolina is consistently in the top-five worst states for motor vehicle fatalities in nearly every category:

- Mileage-based fatality rate – 1st
- DUI fatality rate – 2nd
- Motorcycle fatality rate – 2nd
- Pedalcycle fatality rate – 4th
- Pedestrian fatality rate – 3rd
- Speed-related fatality rate – 2nd
- Population fatality rate – 4th
- Tree-related fatality rate – 1st
- Road departure fatality rate – 3rd
- Fixed-object fatality rate – 1st

South Carolina Strategic Highway Safety Plan

The South Carolina Strategic Highway Safety Plan (SC SHSP) (2015) is the product of a multi-Agency Steering Committee comprising members from the Education, Enforcement, and Engineering sectors (*3*). Although agreeing to utilize a data-driven approach to identify the nine Emphasis Areas, priority categories may have differed among the agencies, because of their respective overall missions. For the South Carolina Department of Public Safety (SCDPS), the focus is centered around enforcement and education, whereas SCDOT is focused on engineering solutions to eliminate roadway fatalities and reduce severe injuries. To achieve the goal of zero traffic fatalities, reductions in the number of fatal and severe injury collisions must be achieved in each of the priority Emphasis Areas. The Emphasis Areas presented in 2015 SHSP were identified using a data-driven process consisting of extensive analysis of fatal and severe injury collision data from 2008 to 2012. In this study the following areas are addressed:

- Roadway departure;
- Age related: young drivers;
- Speed related;
- Vulnerable roadway users: motorcyclists and moped operators; and
- Impaired driving.

Research Method

The overarching goal for this research is to identify proven programmatic strategies that could be used to reduce the number of fatal and serious injury vehicular crashes in South Carolina. Several tasks were undertaken to accomplish the research goals and objectives as follows:

Task A: The research team worked with the steering committee to identify the Emphasis Areas based on the SC SHSP to help direct the safety program research.

Task B: The research team conducted a national search for effective safety programs based on the safety Emphasis Areas identified in the previous task. Research was conducted through a variety of resources, such as National Highway Traffic Safety Administration's (NHTSA's) safety assessments, best-practices safety awards, reviews of the SHSP implementations, commercial literature, peer-reviewed journals, and interviews with DOT personnel.

Task C: The research team conducted in-depth program assessments of successful prior safety program implementations in other states across the nation. A significant amount of information was obtained for each safety program, including: (1) the type of action (engineering, enforcement, education, or other); (2) the parameters of implementation (before and after conditions—crash statistics, demographics, traffic levels, socioeconomics, policy/legislative factors, and environmental factors); (3) cost to implement (low, moderate, and high); (4) the level of effectiveness and expected benefit (low, moderate, high); and (5) any known implementation risks and mitigation factors. The information obtained in this task was used to determine if the potential safety program would likely be successful in South Carolina, and subsequently to determine if the ROI would be an efficient use of resources. Among the programs that were identified in this research, the research team only evaluated programs that were not fully utilized by SCDOT or other stakeholder agencies. The findings presented in this paper all produce a positive benefit cost ratio.

Task D: For each program, the research team investigated potential legislative or policy barriers to successful implementation of the proposed programs in South Carolina. Often times, there were agency policies or state legislation in place that contradicted with new or different programs. In cases where barriers were identified, the research team sought legislation or policy information from the state that successfully implemented the program as an example. While policy and legislative changes may be required for successful implementation of a program, it is understood that this can take a considerable amount of time. Thus, the research team also tried to identify programs that could be implemented with little or no policy or legislative action needed. Unfortunately, many of the most successful programs required legislative or policy action.

Task E: The authors evaluated the benefits and costs of potential safety program implementations to justify the effectiveness of the program in South Carolina. The team estimated the cost-benefit ratio for each program based on information obtained from states that have successfully implemented the program and given the magnitude of the problem in South Carolina. In calculating the actual benefit cost ratio, the 2015 US Department of Transportation fatal crash costs valuation of $9,400,000 was used (*4*). In most cases, benefit evaluations were limited to fatal crashes, but some also included reductions in injuries and PDO crashes. To simplify the ROI comparisons, only fatal crashes are reported.

The following sections present a few of the programs evaluated for the Emphasis Areas in this research. In each section, the description provides a review of conditions in SC, legislative and policy barrier descriptions, and expected impacts of changing the policies or legislation.

Emphasis Area 1: Tree-Related Fatalities (2015: 191 Fatalities, 19.5%)

South Carolina Crash Statistics and Current Policies

South Carolina ranked 1st in the nation for the highest fatality rate (0.32 per 100,000 population) for crashes involving trees. The national average tree-related fatality rate is 0.12 per 100,000 population, thus South Carolina is 165% above the national rate (*5*). Extensive research has been conducted nationally to determine the effect of allowing trees to re-establish in areas that were intended for clear zones. A prior SCDOT research study conducted an analysis of the primary contributing factors in fatal fixed-object crashes both nationally and in South Carolina, and showed a striking difference between the two in terms of the involvement of trees—50% in South Carolina, yet only 21% nationally. Almost 50% of tree-related crashes occurred on secondary roads, with another 25% on primary roads, and approximately 15% on Interstates (*5*). The same study surveyed 131 randomly selected sites to determine if recommended clear zones (or safe recovery areas) were provided. Of these, only 12 sites met the recommended criteria. The research team also analyzed 58 control sites that had no instances of fixed-object crashes within the three-year study period. For these 58 control sites, 47 met the minimum clear zone requirements, and only 11 did not. Using an odds ratio test for this sample, researchers determined that the odds of a site having a fixed-object crash are 42 times higher if the minimum clear zone is not met (*5*).

In 2008, the Federal Highway Administration (FHWA) published *Vegetation Control for Safety* (*6*), providing the following statement: "All trees within the clear zone should be cut while they are still small saplings rather than small trees. At that time, they are easy to cut off at ground level and cause no stump problems. Also, no one will be tempted to try to save a beautiful, but hazardous, tree in the roadway clear zone." This statement succinctly summarizes the problems associated with allowing vegetation regrowth on the roadside. While the FHWA policy sounds straightforward, SCDOT must also contend with Interstate vegetation management laws and local tree ordinances as it strives to provide recommended clear zones on each side of the travel way. The information in this section is offered to illustrate the additional challenge present in establishing a safer roadside environment. The Interstate Vegetation Management Statute (SC Code 57-23-800) requires the SCDDOT to mow and manage vegetation in the medians, roadsides, and interchanges along the Interstate highway system. The statute also stipulates that mowing will not extend more than 30 feet from the edge of the Interstate pavement. There are a few exceptions to this distance requirement for beautification, outdoor advertising, and wildlife management purposes. However, based on the typical Interstate speeds, recovery areas greater than 30 feet and up to 46 feet should be provided.

Some local tree ordinances may also affect the SCDOTs ability to manage roadsides for safety concerns. In South Carolina many of these rights for establishing tree ordinances are protected under the Local Government Comprehensive Planning Enabling Act of 1994, SC Code §6-29. One program called Tree City USA has been facilitating the development of tree ordinances at the local municipality level. Some of the requirements that result from these tree ordinances may stipulate that trees be planted in medians or buffers, or that original wooded areas be preserved. These regulations or even protection status may be based on tree size, type of tree, if the tree is a native species or not, how rare a tree is, or if it has historical significance. In some cases, removal permits may be acquired, but may require replanting and may also require tree surveys by a forester, arborist, or civil engineer. If allowed, some permits can be obtained for lots of trees, whereas others will have to be acquired for each tree individually. The price for obtaining these permits varies as well, and stiff penalties apply if the tree ordinance is not followed.

Review of Successful Safety Programs

Washington, Connecticut, and North Carolina have instituted more rigorous management policies for purposes of providing safer roadsides and to reduce the incidence of tree debris and road blockage during emergency situations such as ice storms and hurricanes. These programs are briefly summarized in the following sections.

The Washington State Department of Transportation (WSDOT) has initiated several programs to deal with the negative effects of trees in the clear zone from crashes, as well as tree impacts from natural disasters. In 2009, King County, Washington had the highest number of single vehicle roadway departure fatal crashes in the state—a total of 76 fatalities. WSDOT instituted a two-phase program: (1) Phase one included installing guide posts and reflectors on existing guardrails and barriers, shoulder and centerline rumble strips, and improving the clear zone, at a cost of $2.2 million; (2) Phase two focused on low-cost improvements, primarily through clear zone reclamation, at a cost of $ 0.6 million (*7*). The reclamation was successful and fatalities dropped significantly following the implementation of the dual-phase project.

When not properly managed, roadside trees can also become susceptible to snow, ice, and wind loads, and can create hazardous conditions for motorists and ultimately cost the DOTs. The State of Washington paid a $10 million settlement to a family because of a fatal tree accident caused by snow loads on a roadside tree. Not only was this loss of life a tragedy, but the costs of storm clean-up and settlements could better be used on proactive management to avoid these incidents.

This project was followed up with the development of the WSDOT Roadside Policy Manual (June 2014), which coordinates with the Integrated Vegetation Management (IVM) Plans and the Management Manual to create a safe, positive environment. Vegetation is intended to have a positive impact on the drivers using the road. The IVM allows vegetation to grow in locations where it is not harmful to motorists; however, removes it from locations where it could cause harm to the traveling public. It also recommends using methods other than chemicals to control vegetation, so as to not negatively affect the surrounding environment. Vegetation is only to be removed as much as necessary for highway purposes (*7*). Similarly, in North Carolina, vegetation management involves replacing the infrangible trees with native grasses and wildflowers. The resulting roadside landscape only requires maintenance every three years to remove unwanted vegetation (*8*).

Connecticut's policy was a direct result of multiple consecutive natural disasters (*9*). After the 2011 Tropical Storm Irene and October Nor'easter caused over $1 billion in damages, Connecticut created a Two Storm Panel to review the storm response to determine what they could have done better and how they could be better prepared in the future. They determined that 90% of the over 800,000 power outages that occurred in both storms were caused by fallen trees. Some residents went without power and clean water for up to 12 days. The storm panel also recommended that the CTDOT increase the budget for tree maintenance to remove potentially dangerous trees more quickly, both dead as well as decaying and dying trees, and trees within the clear zone. Not only does the removal of trees decrease the risk of a fatal or injury crash, it also allows for sunlight to help melt snow and ice during winter storms, reducing the risk of injury and death and decreasing the cost needed for winter maintenance (*10*). The Connecticut General Statute 13a-140 gives authority to the DOT Commissioner with regards to roadside trees. This enables them to cut trees, remove or prune any tree, shrub, or other vegetation located within the highway right-of-way, whenever necessary, for the safe convenient travel by the public, and to reclaim clear zones by removing trees and brush, both healthy and unhealthy, that exist within the "clear zone." As of August 2013, they had trained tree wardens employed in each of four districts. The tree wardens report to a State Licensed Arborist, and they are responsible for identifying dead, dying, and decaying trees and establishing tree management priorities. The budget is about $1.5 million per year, with the four tree wardens.

Summary of Benefits and Recommendations for South Carolina

Recommendations for South Carolina include legislative policy changes to allow SCDOT full control to maintain sufficient road side clear zones within the limits of the right-of-way. A robust vegetation management policy should also be adopted to specify how clear zones can be effectively landscaped and protected from regrowth. Tree wardens could be used to monitor and manage trees in clear zones, as well as ensure that when trees are removed, replacement trees are planted in areas outside the clear area. These policies are proactive, and promote vegetation where it cannot cause harm to motorists and remove it from locations where it can. Proper management and maintenance of trees in clear zones provide significant benefits including: decreased number of injuries and fatalities caused by vehicles hitting trees, improved highway traffic sign visibility, improved sight distance around curves, reduced possibility of trees falling across the roadway during inclement weather, and the re-establishment of drainage ditches. Finally, natural disasters are more easily managed as a result of reduced tree-clearing activities and their associated costs. Several states have been able to strike a balance between the desire for beautification and environmental protection while also keeping roadsides safe for motorists.

Considering the magnitude of the roadside hazard problem, and the deficiency of the clear zones in these areas, it appears that by providing recommended clear zones (or safe recovery areas) for motorists who leave the roadway, South Carolina could realize a notable decrease in roadway fatal and injury crashes. This is particularly significant realizing that many times it is for reasons other than driver error (i.e., blown tire, struck by another vehicle, avoiding an accident, avoiding deer, etc.).

Emphasis Area 2: Young Drivers (2015: 40 Fatalities, 4%)

South Carolina Crash Statistics and Current Policies

In 2015, there were 37 fatal crashes (40 fatalities) among 15–17-year-old drivers, which produced a crash cost of $347.8 million in South Carolina. Young people make up 14.9% of licensed drivers in South Carolina, yet they account for 22.9% of fatal and severe injury crashes (*3*). Although South Carolina does have a Graduated Drivers Licensing (GDL) program in place, the components of the law are some of the most lenient in the country. South Carolina allows young drivers to receive a learners permit at age 15, a restricted license at age 15.5, and a full license at age 16—a full year ahead of the strictest policies. Further, supervised behind-the-wheel experience is also limited, leaving teens to figure things out on their own. On a positive note, South Carolina meets the strictest requirements for night-time driving. However, the limited ability to identify teen drivers on the road after hours may reduce the effectiveness of this provision.

Review of Successful Safety Programs

Along with other safety measures, GDL is the most comprehensive program in place aimed at reducing the number of fatal crashes among young drivers. Increasing the knowledge and awareness of young drivers through educational programs, such as "Alive at 25," also helps to reduce the number of fatal and severe injury crashes. GDL programs have reduced teen crashes by 10–40% on average through programs with the National Institutes of Health, the Insurance Institute for Highway Safety (IIHS), and the NHTSA (*11*). The variation in crash reduction is affected by the specific GDL program parameters in place in each state and the applicable age ranges for program implementations.

The GDL programs include a three-stage criterion for granting young drivers full driver's license privileges. The three stages are: a supervised learning period, a restricted intermediate licensing stage, and a full license stage, in which the driver is granted an unrestricted license after fulfilling all requirements. The purpose of the GDL programs are to maximize experience while minimizing common risks that teens face while driving (*12*). Some of the effective requirements of GDL programs are a minimum age of 15.5 to obtain a learner's permit; a waiting period of at least three months, after obtaining a learner's permit, before applying for an intermediate license; a minimum of 70 hours of supervised driving; a minimum age of at least 16 years to obtain an intermediate license; a minimum age of 17 years to obtain a full license; and night-time driving restrictions. Other restrictions are put into effect that help lower fatal and injury crashes, such as teen drivers and passengers must wear a safety belt, and teen drivers are not allowed to text or talk on the phone while driving.

Studies have shown that U.S. states and Canadian provinces that have put some level of GDL laws into effect have reduced crashes by 20–40%. The stricter GDL laws can reduce fatal crashes by up to 30%. Night-time restriction is the most effective in reducing fatal crashes. In the U.S., enforcing a 9 p.m. driving restriction can reduce fatal crashes by up to 17% (*11*). In California, after the GDL laws went into effect, Thomas Zwicker and his associates used Auto-Regressive Integrated Moving Average time series analysis to study the resulting change. They discovered a 23% decrease in 16-year-olds involved in fatal and injury crashes (*13*).

GDL Calculator by Insurance Institute for Highway Safety. GDL laws have been proven to be effective in preventing fatal crashes; however, GDL laws vary by state. The IIHS created a calculator to determine how much each state could benefit from enforcing the strictest GDL laws. The calculator uses research to determine which aspects of GDL laws are most productive in limiting fatal crashes. The five key provisions included in the model are: the age allowed to get a permit; the number of practice hours required before getting the intermediate license; the minimum age allowed before getting a full license; the night-time driving restrictions; and the allowance of teen passengers. IIHS studied the rate of fatal crashes per 100,000 teenage drivers in every state and used the GDL laws in all of the states to determine how much each provision of the GDL law would reduce the rate of fatal crashes.

There are a few states that are touted as having the strictest GDL laws in the country including Connecticut, District of Columbia, and New Jersey (*14*). Connecticut's laws on passengers and permit age are the strictest that have been adopted. Connecticut requires parents or guardians of drivers under the age of 18 to attend a 2-hour class that explains the laws for teen drivers and related issues, before the teen can take the test to get their intermediate license. The District of Columbia follows the strictest provisions for permit age and the number of teen passengers. However, they require everyone go through the intermediate stage until they are 21 years old. They also require drivers under 18 to obey the laws regarding night-time driving and passengers, even if they have their full license. Still yet, IIHS estimates that the District of Columbia can further achieve a 17% reduction in fatal crashes and a 14% reduction in collision crashes (*15*). New Jersey has a law that requires all teenagers on a learner's permit and all drivers on probation to have a decal on their license plate, so the police can more easily identify if a law is being broken—especially at night (*16*). New Jersey could also have an additional 25% reduction in fatal crashes if they enacted the strictest laws (*15*).

Summary of Benefits and Recommendations for South Carolina

Although South Carolina does have an existing GDL program in place, because of the leniency of the program it does not have achieved full potential in reducing the number of fatalities. For South Carolina, the IIHS GDL Calculator estimates a 45% reduction in fatal crashes and a 22% reduction in collision claims with improvements in the GDL laws. The reduction could be realized by adopting three stricter GDL criteria, including: raising the permit age to 16 and the unrestricted licensing age to 17; raising the minimum number of practice hours to 70; and restricting teen passengers during the intermediate driving phase. Roughly 17 teen drivers could be saved each year with these GDL policy improvements in

South Carolina. Further, 22% of current collision claims would also be reduced for this group of young drivers, which would lead to fewer injuries and property losses—a significant economic and safety benefit for the state. These changes are highly recommended and merely tweak existing policies. Further recommendations include making night-time restriction more effective by utilizing decals on teen drivers' license plates for easy identification by the police officers. Finally, parents should be required to undergo training and a test to ensure that they understand the laws for which the teens will be held accountable.

Emphasis Area 3: Impaired Driving (2015: 301 Fatalities, 30.7%)

South Carolina Crash Statistics and Current Policies

In 2015, South Carolina had 301 fatalities (30.7% of total fatalities) involving a driver with blood alcohol content (BAC) over 0.08, which is considered legally drunk. This is the highest percentage in the country. South Carolina also has some of the weakest laws in the United States relating to DUI offenders. A driver's license is suspended for refusal to take a test for alcohol; however, a driver may obtain a temporary license or a route-restricted license on release from jail. Emma's Law, passed in 2014, increased the penalties for DUI convictions, requiring first-time convicted offenders with a BAC of 0.15% or greater to complete the state's Ignition Interlock Device Program. The same law removed the hard suspension period for second and subsequent convictions, allowing drivers to get their licenses back sooner by completing an Alcohol and Drug Safety Action Program. A 2016 Impaired Driving Assessment refers to this SC statue as "...poorly drafted and archaic...and mandates unsafe roadside practices endangering the public and the officer making the stop" (*17*).

Recent data released by the SCDPS show a significant number of DUI arrests are made in the state each year. However, because of legislative loopholes in DUI statutes, prosecution has become difficult. Another recommendation proposes increasing the number of solicitors to handle DUI cases. South Carolina is one of only two states in the nation where police officers prosecute their own DUI cases. Solicitors could assist officers, particularly in cases held in higher court, by making appropriate motions and responding to motions made by the defense. A short-term alternative to solicitors might be resource prosecutors who train officers on best practices for DUI prosecution such as those used by The Kansas Traffic Safety Resource Prosecutors (*18*). By comparison, Nebraska has one of the strongest laws for prosecuting DUI offenders. Administrative licenses are suspended for 180 days and only given back if an ignition interlock system is installed. They also require first-time offenders and repeat offenders to install an ignition interlock system, regardless of BAC content (*19*).

In South Carolina from 2008 to 2012, drivers ages 25–34 had the highest number of DUI incidents and the highest percentage of crashes out of any age group. There were 1,576 incidents of DUI among 15–19-year-old drivers, all of which are illegal under state law. By gender, males had the highest incidence of driving while intoxicated. In a five-year span, South Carolina had over 900 people with at least two DUI offenses. Many programs could either be implemented or enforced better, such as zero tolerance, DUI/DWI courts, etc., that could reduce all of the DUI-related factors in South Carolina (*20*). South Carolina's most recent Impaired Driving Assessment lends further support to this statement, citing Alli's Law—proposed legislation that would require responsible beverage server training as a condition of business liquor licensure (*17*).

Review of Successful Safety Programs

Arizona enacted the use of ignition interlock systems whenever a driver is charged for DUI. Before this program was enacted in 2007, DUI fatalities comprised 54% of the total fatalities in Arizona. By 2012, DUI fatalities had dropped to 28% of the total fatalities. Understandably, ignition interlock systems have been approved in some form of legislation in every state because they offer an effective method of preventing drunk driving. They also have a relatively low cost, averaging from $3 to $7 a day depending on the type of data and monitoring that is done. Further, the devices are tamper resistant and monitor almost everything about the vehicle and driver (*21*).

Georgia DOT established DUI courts in 2003, and the goal of these courts is to ensure abstinence from drugs and alcohol for all participants. The courts that were established included: a judge; a DUI court coordinator; and a case management clerk. The majority of DUI court participants (95%) were addicted to alcohol and have had at least three lifetime DUIs. Over the four-year period, there was a 79% participant retention rate. This indicates that once in the program, most participants had a motivation to complete it. Georgia now has 19 DUI courts. Graduates of the Georgia DUI court program are four times less likely to have a repeat DUI arrest and 20% less likely to be arrested for a felony within 24 months after graduation from the program (*22*).

Summary of Benefits and Recommendations for South Carolina

As mentioned earlier in previous sections, South Carolina has one of the weakest laws in the U.S. relating to DUI offenders. An assessment of the current law refers to it as "poorly drafted and archaic" (*17*). South Carolina is one of only two states in the nation where police officers prosecute their own DUI cases. South Carolina police officers are not trained as lawyers, yet they have to go toe-to-toe with practicing lawyers to get a conviction. While the officers are making arrests, it is not common to see retribution in these cases (*17*). Based on the review of the states with successful policies for reducing DUI-related fatalities, it is recommended that South Carolina adopt the use of solicitors to engage in court proceedings, ignition interlock systems for all drivers who are charged for driving under influence, and, if possible, DUI court processes. Arizona has experienced a 28% drop in total fatalities after adopting ignition interlocks (*21*), and Georgia found that drivers completing the DUI court program are four times less likely to have a repeat DUI arrest and 20% less likely to be arrested for a felony within 24 months after graduation from the program (*22*).

Emphasis Area 4: Speed-Related Fatalities (2015: 361 Fatalities, 36.9%)

South Carolina Crash Statistics and Current Policies

In 2015, almost 40% of the fatalities on South Carolina roadways were speed related. Speed related is defined as: exceeding authorized speed limit, and/or driving too fast for conditions. This is the second highest rate in the nation. Speed is a behavioral issue that must be managed through enforcement activities, and the reality is that drivers far outnumber enforcement officers. Auto-enforcement with cameras has the potential to expand speed management programs and reduce crashes—not only speed-related crashes, but all crashes. Speed cameras work by using a photo radar to determine when a vehicle exceeds a set limit, and then it takes a picture of the vehicle and license plate to enable vehicle identification for ticketing purposes (*23*).

In 2010, South Carolina Governor Mark Sanford signed a law effectively banning the use of red light cameras and speed cameras in the state. The legislation was unanimously approved in both the South Carolina House and Senate. The law does allow photo ticket use during emergencies declared by the governor or president; however, the ticket must be personally delivered by an officer within 1 hour, which is not practical for most situations and renders photo ticketing useless.

Review of Successful Safety Programs

Camera enforcement began for the first time in 1987 in Arizona. Although several states have banned photo enforcement, over a dozen others allow its use and have found it to be quite effective. Since the late 1980s, speed cameras have been used in 12 states and the District of Columbia and have lowered fatal crashes up to a 25% at fixed camera sites and up to almost 50% with mobile camera operations (*24*). Speed cameras also allow for better traffic flow (*25*). Figure 1 shows southeastern states that are using photo enforcement for speed and/or red light running enforcement (*23*).

Tennessee and Alabama are two states that utilize both red light running cameras and speed cameras. They put cameras in critical speed locations such at school zones and sharp curve areas. The purpose of putting photo enforcement on sharp curves is to prevent speeding on roads where it is difficult to see the entire stretch of road. In a study of the effectiveness of the speed cameras in Chattanooga, the drivers were fined if they exceeded the speed limit. Both citations and fatal and injury crashes decreased. Between 2000–2001, there were five deaths and 32 injuries on S curves; however, after camera installation (2004–2005), there were no deaths and only 12 injuries.

Another method of photo enforcement is through mobile vans. The City of Chattanooga has two vans that carry speed cameras, which operate in the same manner as the fixed cameras. They allow for a wider range of operation because they can be moved wherever there is a need for them (*25*). The number of citations issued by the automated system far exceed that which can be written by hand in the field by officers. Further, automated enforcement would allow scarce police resources to be used more effectively for other important tasks that cannot be handled by a computer system.

Montgomery County in Maryland is another example of success by speed cameras. By 2014, they were using 56 fixed cameras, 30 portable cameras, and six mobile vans. All of the cameras are used in either residential areas or school zones. There is a $40 fine for the citation and no points on the license. Within six months of the implementation of the speed cameras, the proportion of drivers traveling ten miles over the speed limit or higher dropped significantly on streets with speed cameras. Seven years after the program was implemented, there is a 59% reduction in the chance that a driver will drive ten or more miles over the speed limit. There was also a 19% reduction in the probability that a fatality or severe injury would occur as a result of a crash on roads with speed cameras.

In 2012, Montgomery County found that implementation of speed-camera corridors reduced the likelihood of crashes even more. Instead of using fixed camera locations, the cameras on the speed-camera corridor were moved around within the corridor, so motorists did not know the specific location within the corridor where speeds were monitored. The resulting effect was that

Figure 1. Southeastern states with red light and/or speed cameras.

drivers reduced their speed for the entire length of road because they were uncertain where the cameras would be operating. The speed-camera corridor approach helped reduce the likelihood of fatal or severe injury crashes by an additional 30% beyond the original 19% (*26*).

Summary of Benefits and Recommendations for South Carolina

The review of successful programs revealed that installation of red light running and speed cameras can significantly reduce the number of deaths and injuries. The analysis of the crashes before and after the installation of speed cameras in Chattanooga showed 100% reduction in deaths on S curves. The Maryland study showed a 59% reduction in the probability of occurring fatal and severe injury crashes. However, South Carolina is far behind adapting these programs as it has effectively banned the use of red light and speed cameras in the state. Knowing the fact that 40% of fatalities are speed related, it is highly recommended that South Carolina should remove the ban and start installing the speed and red light running cameras at the problematic locations.

Emphasis Area 5: Vulnerable Road Users—Motorcyclists (2015: 135 Fatalities, 13.8%)

South Carolina Crash Statistics and Current Policies

A total of 135 motorcyclists were killed in South Carolina in 2015, which is 13.8% of all occupants killed that year. Overall, 74% of South Carolina motorcycle riders killed were not wearing a helmet (*1*). South Carolina ranks second in nation in motorcycle fatalities with 3.8 fatalities per 100,000 population, following Wyoming at 4.2 fatalities per 100,000 population. (Note: The national statistics include both motorcycle and moped riders.)

In 1967, states were required to enact helmet laws in order to qualify for certain federal safety programs and highway construction funds. South Carolina enacted the universal helmet law on July 1, 1967. However, in 1976, several states successfully lobbied Congress to stop the USDOT from assessing financial penalties on states without the universal helmet law. On June 16, 1980, the South Carolina legislature had reduced the helmet law coverage to those 20 and younger. A decade later, Sosin et al. (1990) conducted a study of motorcycle fatalities from 1979 through 1986 (*27*). They found that over half of the motorcycle fatalities were associated with head injuries, and rates per population for motorcycle fatalities associated with head injury (adjusted by age, sex, and race) were almost twice as high in states without universal helmet laws as in states with universal helmet laws. In the two states that dropped universal coverage during the study period, motorcyclist fatalities per population rose substantially: by 184% in South Carolina and by 73% in Wyoming.

Review of Successful Safety Programs

On July 1, 2000, Florida repealed the legal requirement that all motorcyclists wear protective helmets. Similar to South Carolina, the helmet law Florida adopted only requires riders aged 20 and younger to wear helmets. However, the law also requires older riders who do not have a minimum of $10,000 medical insurance coverage to wear helmets. NHTSA Traffic Safety Facts (2005) indicate a substantial increase in motorcyclists killed in Florida beginning in the first six months of 2000 (the repeal of the all rider helmet law went into effect on July 1, 2000). Fatalities in the two years following the law change (2001–2002, $N = 575$) were 71% greater than the two years before the law change. This is almost double the increase in motorcycle fatalities for the nation (*28*).

Currently in the southeast, only South Carolina and Florida maintain partial laws, while all other states have universal laws. Louisiana has one of the most recent (2004) reinstated universal helmet laws which require all motorcyclists (both riders and passengers) to wear helmets at all times. Numerous data analysis techniques were conducted to compare helmet use and injury levels for before and after the law change. The analysis showed a strong positive effect of helmet use in post-law reinstatement. The odds of wearing a helmet in a crash after the law was reinstated were 11.7 times greater compared with wearing a helmet during the pre-law time period. The injuries were appreciably lower in the post-law period compared with the pre-law period. Fatalities were 30% less likely during the post-law period as compared with pre-law period. Moreover, the analyses also showed that there were also fewer severe and fatal crashes following the law change. When the universal helmet law was reinstated, Louisiana experienced its first decline in motorcyclist fatalities in six years. The results clearly support that the decrease in fatal motorcycle crashes was directly affected by the helmet law reinstatement in Louisiana (*29*).

According to results from the National Occupant Protection Use Survey, the overall rate of DOT-compliant motorcycle helmet use in the United States was 60% in 2013 (*30*). Helmet use continued to be significantly higher (approaching 90%) in States that required all motorcyclists to be helmeted than in other States. In States without universal helmet laws, 59% of motorcyclists killed in 2013 were not wearing helmets, as compared with 8% in States with universal helmet laws. Per vehicle mile traveled in 2013, motorcyclist fatalities occurred 26 times more frequently than passenger car occupant fatalities in motor vehicle traffic crashes. NHTSA estimates that helmets saved the lives of 1,630 motorcyclists in 2013. If all motorcyclists had worn helmets, an additional 715 lives could have been saved. Helmets are estimated to be 37% effective in preventing fatal injuries to motorcycle riders and 41% for motorcycle passengers (*31*). In other words, for every 100 motorcycle riders killed in crashes while not wearing helmets, 37 of them could have been saved had all 100 worn helmets.

Summary of Benefits and Recommendations for South Carolina

In 2005, the state legislature voted on a bill that included both a primary safety belt and a primary helmet laws. Although the safety belt law passed, the helmet law did not. Many point to the powerful motorcycle lobby groups as the reason for the demise. Motorcycle lobby groups give utmost importance to their personal liberty, and simply ignore the detrimental effect of not wearing a standard motorcycle helmet. SCDOT has been trying to reinstate the universal helmet law in South Carolina for years; in light of the research on helmet effectiveness and the review on the states who are successful in reducing fatalities related to helmet use, it is recommended for South Carolina to reinstate the universal helmet law sooner than later.

Return on Investment

In this section the ROI for each emphasis area is presented. The ROI is calculated for the year 2015 using a USDOT fatal crash cost valuation of $9,400,000 (*4*). Table 1 contains the summary table on ROI.

Conclusions

While SCDOT has a federal requirement to develop and maintain the SHSP which identifies the state's key transportation safety needs and guides investment decisions toward strategies and countermeasure with the most potential to save lives and prevent injuries, South Carolina legislation and state policies have effectively blocked many paths to safety improvements. To make significant gains, all stakeholder partners (including legislative and policy makers) must have a complete vision

Table 1. Return on Investment Summary

Treatment	Number of fatalities	Total fatal crash cost (in millions of $)	Fatal crash/ fatality reduction percentage	Estimated lives saved	Crash cost saving (in millions of $)	Program implementation cost (in millions of $)	B:C ratio
Clear zone reclamation	191	$1,795	27%	52	$454	$15.75	29:1
Universal helmet law	135	$1,269	24%	32	$302	$1.50	202:1
Stricter GDL	40	$348	45%	18	$157	$ 0.00	156:1
Camera speed enforcement	50	$470	19%	10	$89	$6.75	13:1
DUI Courts	301	$2,829	1.70%	6	$48	$1.50	32:1

Assumptions for clear zone reclamation:

1. There is only a reduction in fatalities.
2. The re-establishment of vegetation in the clear zone is less than 50% of the intended clear zone distance; therefore, a 27% reduction in fatal crashes can be obtained for reclaiming up to 50% of the intended clear zone.
3. This estimate does not consider the potential benefits of decreased tree removal and reduced hazards accrued during natural disasters.
4. The cost of clearing was taken from a recent Interstate-26 clearing project, where 56 acres were cleared for $5 million. Therefore, clearing is estimated at $11,000 per linear foot-mile with traffic control.
5. Treatment sites are estimated as one mile in length.
6. The average clear zone regrowth distance was taken from the FHWA-SC-09-01 as shown in Table 6.3.
7. Cost to clear = Average clear zone regrowth * Number of sites * $11,000 per linear foot-mile.
8. Crash saving benefits = 2015 fatal crashes involving trees * $9.4 million fatal crash costs * 27% reduction across all road types.
9. B:C Ratio = Crash saving benefits / Cost to Clear.

Assumptions for universal helmet law:

1. There is only a reduction in fatalities. Although some fatalities may actually convert to injuries, this estimate is conservative because no crash savings for reductions in injury crashes were considered.
2. The reduction in fatal crashes with universal helmet law implementation is 37%.
3. In South Carolina 74% do not wear a helmet, and would be affected by the universal helmet law.
4. Based on national statistics, the universal helmet law would induce an 87% compliance rate with helmet usage.
5. The total reduction 37%*74%*87% = 24%.

Assumptions for stricter GDL:

1. Using the estimate of fatal crash reductions from the IIHS GDL Calculator, a stricter GDL program could lead to a 45% reduction in fatal crashes. To maintain units of fatalities, a standard conversion factor, 1.07 fatalities per fatal crash, was assumed. So for 37 fatal crashes, the number of fatalities was = 37*1.07 = 40
2. Three provisions would be adopted in full including: raising the permit age to 16, raising the unrestricted licensing age to 17, raising the minimum number of practice hours to 70, and restricting teen passengers during the intermediate driving phase.
3. There is only a reduction in fatalities. Although some fatalities may actually convert to injuries, this estimate is conservative because no crash savings for reductions in injury crashes were considered.

Assumptions for camera speed enforcement:

1. There is only a reduction in fatalities. Although some fatalities may actually convert to injuries, this estimate is conservative because no crash savings for reductions in injury crashes were considered. The literature indicates a 12% overall crash reduction with the use of speed cameras.
2. The reduction in fatal crashes with speed-camera implementation is estimated at 19%.
3. The cameras cost about $50,000 to purchase and $25,000 to install. Monthly operating costs are about $5,000 (per camera system).
4. Each fatal crash site is an individual site.
5. 50 cameras will be purchased to install at 50 fixed sites in close proximity to 50 fatal crash sites.

Assumptions for DUI courts:

1. The result of the programs that are implemented is a 1.7% reduction in DUI crashes (GA DUI Courts). To maintain units of fatalities, a standard conversion factor, 1.07 fatalities per fatal crash, was assumed.
2. The cost of the program is $1,500,000 for five courts (GA DUI Courts).

of their role and responsibilities in the priority programs and seek collective efficacy through collaboration across the state. The development of this comprehensive safety program assessment, along with identification of funding sources, will enable forward movement on all fronts. Using a data-driven approach to safety program selection will yield support for changes in programs, policies, and standards, and have positive impacts on safety, operational, and economic aspects of the South Carolina roadway system. Further, the implementation of a data-driven safety management program will help to assure that the most appropriate strategies are implemented.

The successful implementation of this research will likely result in a substantial reduction in loss of life and injuries associated with motor vehicle crashes in the state of South Carolina. This research investigated the total potential for fatal crash reduction given adoption of safety programs that proved to be successful in other states. If the policies discussed in this research were implemented at the assumed levels in South Carolina, on an annual basis, 118 fatalities could potentially be reduced. That means of the total 979 fatalities, 12.05% of lives could be saved. From a matter of priority, the number of lives saved would be a more meaningful prioritization metric than dollars saved. Therefore, prioritizing based on the number of potential lives saved each year would allow program selection to get the most positive benefits. Also each of the programs studied in this research has a significant net positive cost ratio, so the state should not go wrong in implementing any of the recommendations. The benefits of these findings are numerous. In addition to reducing the number of crashes and the resulting deaths and injuries, South Carolina will benefit from improved system operations and reduced delay, and decreased fuel consumption and emissions, as well as potential cost savings for SCDOT and other stakeholder agencies.

In terms of transferability of this research, the findings from this study can also be used for safety improvement decisions in other states. If a state has identified legislative barriers similar to South Carolina, the authors have summarized some successful examples of state policies that have been adopted which significantly reduced the number of fatal crashes. Moreover, a detailed process for calculating the ROI of each of these policies has also been summarized.

Author Contributions

The authors confirm contribution to the paper as follows: study conception and design: South Carolina Department of Transportation and Jennifer H. Ogle; data collection: Jennifer H. Ogle, Sababa Islam, Kweku Brown, William J. Davis, and Wayne A. Sarasua; analysis and interpretation of results: Jennifer H. Ogle, Sababa Islam, Kweku Brown, William J. Davis, and Wayne A. Sarasua; draft manuscript preparation: Primary – Jennifer H. Ogle and Sababa Islam, Contributors – Kweku Brown, William J. Davis, and Wayne A. Sarasua. All authors reviewed the results and approved the final version of the manuscript.

References

1. South Carolina Department of Public Safety. *South Carolina Traffic Collision Fact Book*. South Carolina Department of Public Safety, 2015. http://www.scdps.gov/ohsjp/fact%20book/2015%20Fact%20Book.pdf. Accessed November 15, 2017.
2. National Highway Traffic Safety Administration. *Traffic Safety Facts 2015*. DOT HS 812 384. U.S. Department of Transportation, National Highway Traffic Safety Administration, 2016. https://www.google.com/url?sa=t&rct=j&q=&esrc=s&source=web&cd=1&cad=rja&uact=8&ved=0ahUKEwi_796d8qrVAhWJWCYKHY28BfQQFggpMAA&url=https%3A%2F%2Fcrashstats.nhtsa.dot.gov%2FApi%2FPublic%2FPublication%2F812384&usg=AFQjCNFO2Y1ngW38OmOtsDEj93iedMKYhg. Accessed July 31, 2017.
3. South Carolina Department of Transportation. *South Carolina's Strategic Highway Safety Plan*. South Carolina Department of Transportation, 2015. http://www.scdps.gov/docs/Target%20Zero_Final_w_Signatures_15APR15.pdf. Accessed July 31, 2017.
4. U.S. Department of Transportation. *Guidance on Treatment of the Economic Value of a Statistical Life (VSL)*. U.S. Department of Transportation Analysis, U.S. Department of Transportation, 2016. https://cms.dot.gov/sites/dot.gov/files/docs/2016%20Revised%20Value%20of%20a%20Statistical%20Life%20Guidance.pdf. Accessed July 31, 2017.
5. Ogle, J., W. Sarasua, J. Dillon, V. Bendigieri, S. Anekar, and P. Alluri. *FHWA-SC-09-01 Report: Support for the Elimination of Roadside Hazards: Evaluating Roadside Collision Data and Clear Zone Requirements*. Transportation Research Board of the National Academies, Washington, D.C., 2009.
6. Eck, R., and H. McGee. *Vegetation Control for Safety*. Federal Highway Administration, 2008 https://safety.fhwa.dot.gov/local_rural/training/fhwasa07018/. Accessed July 31, 2017
7. Washington State Department of Transportation. *Roadside Policy Manual*, M 3110.02. Washington State Department of Transportation, 2014.
8. *North Carolina Department of Transportation's Tree Planting Initiatives*. North Carolina Department of Transportation, 2012. http://www.dcvb-nc.com/ScenicNC/2012_NCDOT_Tree_Planting_Overview.pdf. Accessed July 31, 2017.
9. Connecting Western Washington. *State to Pay Family $10M over Fatal Tree Incident*, 2012. http://www.king5.com/news/local/state-to-pay-family-10m-over-fatal-tree-incident/246987819. Accessed July 31, 2017.
10. *Connecticut Department of Transportation Roadside Tree Removal Practices*. Connecticut Department of Transportation, 2013. http://www.ct.gov/dot/cwp/view.asp?A=2137&Q=529766. Accessed July 31, 2017.
11. McCartt, A. T., E. R. Teoh, M. Fields, K. A. Braitman, and L. A. Hellinga. Graduated Licensing Laws and Fatal Crashes of Teenage Drivers: A National Study. *Traffic Injury Prevention*, Vol. 11, No. 3, 2010, pp. 240–248.
12. *Teen Driving Safety Leadership Award, National Safety Council Mission*. http://www.nsc.org/learn/NSC-Initiatives/Pages/teen-driving-whats-working.aspx. Accessed November 15, 2017.
13. Zwicker, T. J., A. F. Williams, N. K. Chaudhary, and C. M. Farmer. *Evaluation of California's Graduated Licensing System*. Insurance Institute for Highway Safety, Arlington, VA, 2006.
14. Insurance Institute for Highway Safety. *An Easy Win-Strong GDL Laws Maximize Benefits*. Insurance Institute

for Highway Safety, Highway Loss Data Institute, 2015. http://www.iihs.org/iihs/sr/statusreport/article/50/3/1. Accessed July 31, 2017.

15. GDL Crash Reduction Calculator. *Insurance Institute for Highway Safety*. Highway Loss Data Institute, 2015. http://www.iihs.org/iihs/topics/laws/gdl_calculator?state = SC. Accessed July 31, 2017.
16. McCartt, A. T., N. J. Oesch, A. F. Williams, and T. C. Powell. New Jersey's License Plate Decal Requirement for Graduated Driver Licenses: Attitudes of Parents and Teenagers, Observed Decal Use, and Citations for Teenage Driving Violations. *Traffic Injury Prevention*, Vol. 14, No. 3, 2013, pp. 244–258.
17. Cynthia, B., L. Chezem, R. Lillis, M. Stout, and T. Woodward. *State of South Carolina: Impaired Driving Program Assessment, Chapter III*. National Highway Traffic Safety Administration (NHSTA), Washington, D.C., 2016, p9 and p49.
18. *Kansas Traffic Safety Resource Prosecutor*, https://www.ktsro.org/kansas-traffic-safety-resource-prosecutor. Accessed February 23, 2018.
19. Impaired, Driving, and State Laws. *Insurance Institute for Highway Safety*. Highway Loss Data Institute, 2017. http://www.iihs.org/iihs/topics/laws/dui. Accessed July 31, 2017.
20. Brown, K., W. A. Sarasua, and J. H. Ogle. Safety Planning: Analysis of the Socio-Economic and Demographic Characteristics of At-Risk Driver Residential Areas in South Carolina. Presented at 95th Annual Meeting of the Transportation Research Board, Washington, D.C., 2016.
21. *Arizona DUI Laws & Penalties*. http://www.duiprocess.com/state/AZ/duilaws.php. Accessed July 31, 2017.
22. *Research Results: Georgia's DUI Courts work*! Judicial Council of Georgia. http://www.chathamcourts.org/Portals/ChathamCourts/State%20Court/Forms/Civil/DUI-Court-Recidivism.pdf. Accessed July 31, 2017.
23. Insurance Institute for Highway Safety. *Speed, State Laws – Automated Enforcement*. Insurance Institute for Highway Safety, Highway Loss Data Institute, 2017. http://www.iihs.org/iihs/topics/laws/automated_enforcement?topicName = speed. Accessed July 31, 2017.
24. Decina, L. E., L. Thomas, R. Srinivasan, L. K. Staplin, and L. L. C. TransAnalytics *Automated Enforcement: A Compendium of Worldwide Evaluations of Results*. No. HS-810 763. Highway Safety Research Center, University of North Carolina, Chapel Hill, NC, 2007.
25. Winkle, J. *Chattanooga's Photo Speed Enforcement Program*. Presented at 87th Annual Meeting of the Transportation Research Board, Washington, D.C., 2008.
26. Hu, W., and A. T. McCartt. Effects of Automated Speed Enforcement in Montgomery County, Maryland, on Vehicle Speeds, Public Opinion, and Crashes. *Traffic Injury Prevention*, Vol. 17, 2016, pp. 53–58.
27. Sosin, D. M., J. J. Sacks, and P. Holmgreen. Head Injury—Associated Deaths from Motorcycle Crashes: Relationship to Helmet-Use Laws. *JAMA*, Vol. 264, No. 18, 1990, pp. 2395–2399.
28. National Highway Traffic Safety Administration. *Traffic Safety Facts 2005*. U.S. Department of Transportation, National Highway Traffic Safety Administration, National Center for Statistics and Analysis. https://crashstats.nhtsa.dot.gov/Api/Public/ViewPublication/810631. Accessed July 31, 2017.
29. National Highway Traffic Safety Administration. *Evaluation of the Reinstatement of the Motorcycle Helmet Law in Louisiana*. Publication No. 346. US Department of Transportation, 2008. https://crashstats.nhtsa.dot.gov/Api/Public/ViewPublication/810631. Accessed July 31, 2017.
30. National Highway Traffic Safety Administration. *Motorcycle Helmet Use in 2013—Overall Results*. U.S. Department of Transportation, National Highway Traffic Safety Administration, 2014. https://crashstats.nhtsa.dot.gov/Api/Public/ViewPublication/812010. Accessed July 31, 2017.
31. National Center for Statistics and Analysis. *Lives and Costs Saved by Motorcycle Helmets*, 2014. National Center for Statistics and Analysis. Report No. DOT HS 812 295. 2016, July. https://crashstats.nhtsa.dot.gov/Api/Public/Publication/812295. Accessed July 31, 2017.

The Standing Committee on Transportation Safety Management (ANB10) peer-reviewed this paper (18-06569).